**STOCKPORT** *Libraries, Advice and Information*
METROPOLITAN BOROUGH COUNCIL

D1757500

CLL 8/12

5 - DEC 2013

Please return/renew this item by the last date shown.
Books may also be renewed by phone or the Internet

TEL: 0161 217 6009
www.stockport.gov.uk/libraries

C2000 00035 7855

The
Triumph Tiger Cub
Bible

Other Veloce publications -

Colour Family Album Series
Alfa Romeo by Andrea & David Sparrow
Bubblecars & Microcars by Andrea & David Sparrow
Bubblecars & Microcars, More by Andrea & David Sparrow
Citroën 2CV by Andrea & David Sparrow
Citroën DS by Andrea & David Sparrow
Custom VWs by Andrea & David Sparrow
Fiat & Abarth 500 & 600 by Andrea & David Sparrow
Lambretta by Andrea & David Sparrow
Mini & Mini Cooper by Andrea & David Sparrow
Motor Scooters by Andrea & David Sparrow
Porsche by Andrea & David Sparrow
Triumph Sportscars by Andrea & David Sparrow
Vespa by Andrea & David Sparrow
VW Beetle by Andrea & David Sparrow
VW Bus, Camper, Van & Pick-up by Andrea & David Sparrow

SpeedPro Series
How to Blueprint & Build a 4-Cylinder Engine Short Block for High Performance by Des Hammill
How to Build a V8 Engine Short Block for High Performance by Des Hammill
How to Build a Fast Road Car by Daniel Stapleton
How to Build & Modify Sportscar/Kitcar Suspension & Brakes by Des Hammill
How to Build & Modify SU Carburettors for High Performance by Des Hammill
How to Build & Power Tune Weber DCOE & Dellorto DHLA Carburetors Second Edition by Des Hammill
How to Build & Power Tune Harley-Davidson Evolution Engines by Des Hammill
How to Build & Power Tune Distributor-type Ignition Systems by Des Hammill
How to Build, Modify & Power Tune Cylinder Heads Second Edition by Peter Burgess
How to Choose Camshafts & Time them for Maximum Power by Des Hammill
How to Give your MGB V8 Power Updated & Revised Edition by Roger Williams
How to Improve the MGB, MGC & MGB V8 by Roger Williams
How to Power Tune the BMC 998cc A-Series Engine by Des Hammill
How to Power Tune BMC/Rover 1275cc A-Series Engines by Des Hammill
How to Power Tune the MGB 4-Cylinder Engine by Peter Burgess
How to Power Tune the MG Midget & Austin-Healey Sprite Updated Edition by Daniel Stapleton
How to Power Tune Alfa Romeo Twin Cam Engines by Jim Kartalamakis
How to Power Tune Ford SOHC 'Pinto' & Sierra Cosworth DOHC Engines by Des Hammill

General
Alfa Romeo Berlinas (saloons/sedans) by John Tipler
Alfa Romeo Giulia Coupé GT & GTA by John Tipler
Automotive Mascots by David Kay & Lynda Springate
British Cars, The Complete Catalogue of 1895-1975 by Culshaw & Horrobin
British Trailer Caravans & their Manufacturers 1919-1959 by Andrew Jenkinson
British Trailer Caravans & their Manufacturers from 1960 by Andrew Jenkinson
Bugatti Type 40 by Barrie Price
Bugatti 46/50 - The Big Bugattis 2nd Edition by Barrie Price
Bugatti 57 - The Last French Bugatti by Barrie Price
Chrysler 300 - America's Most Powerful Car by Robert Ackerson
Cobra - The Real Thing! by Trevor Legate
Cortina - Ford's Bestseller by Graham Robson
Daimler SP250 'Dart' by Brian Long
Datsun/Nissan 280ZX & 300ZX by Brian Long
Datsun Z - From Fairlady to 280Z by Brian Long
Dune Buggy Handbook by James Hale
Fiat & Abarth 124 Spider & Coupé by John Tipler
Fiat & Abarth 500 & 600 by Malcolm Bobbitt
Ford F100/F150 Pick-up by Robert Ackerson
Grey Guide, The by Dave Thornton
Jim Redman - Six Times World Motorcycle Champion by Jim Redman
Lea-Francis Story, The by Barrie Price
Lola - The Illustrated History (1957-1977) by John Starkey
Lola T70 - The Racing History & Individual Chassis Record New Edition by John Starkey
Mazda MX-5/Miata 1.6 Enthusiast's Workshop Manual by Rod Grainger & Pete Shoemark
Mazda MX-5/Miata 1.8 Enthusiast's Workshop Manual by Rod Grainger & Pete Shoemark
Mazda MX-5 Renaissance Sportscar by Brian Long
MGA by John Price Williams
Mini Cooper - The Real Thing! by John Tipler
Motor Museums of the British Isles & Republic of Ireland by David Burke & Tom Price
Porsche 356 by Brian Long
Porsche 911R, RS & RSR 4th Edition by John Starkey
Porsche 914 & 914-6 by Brian Long
Prince & I by Princess Ceril Birabongse
Rolls-Royce Silver Shadow/Bentley T Series Corniche & Camargue Updated Edition by Malcolm Bobbitt
Rolls-Royce Silver Spirit, Silver Spur & Bentley Mulsanne by Malcolm Bobbitt
Rolls-Royce Silver Wraith, Dawn & Cloud/Bentley MkVI, R & S Series by Martyn Nutland
Singer Story: Cars, Commercial Vehicles, Bicycles & Motorcycles by Kevin Atkinson
Taxi! The Story of the 'London' Taxicab by Malcolm Bobbitt
Triumph Motorcycles & the Meriden Factory by Hughie Hancox
Triumph TR6 2nd Edition by William Kimberley
Velocette Motorcycles - MSS to Thruxton by Rod Burris
Volkswagen Karmann Ghia by Malcolm Bobbitt
Volkswagens of the World by Simon Glen
VW Bus, Camper, Van, Pickup by Malcolm Bobbitt
Works Rally Mechanic by Brian Moylan

First published in 2000 by Veloce Publishing Plc., 33, Trinity Street, Dorchester DT1 1TT, England. Fax: 01305 268864/e-mail: veloce@veloce.co.uk/ website: www.veloce.co.uk

ISBN: 1 901295 44 3/UPC: 36847-00144-5

© 2000 Mike Estall and Veloce Publishing Plc
All rights reserved. With the exception of quoting brief passages for the purpose of review, no part of this publication may be recorded, reproduced or transmitted by any means, including photocopying, without the written permission of Veloce Publishing Plc.
Throughout this book logos, model names and designations, etc., may have been used for the purposes of identification, illustration and decoration. Such names are the property of the trademark holder as this is not an official publication.
Readers with ideas for automotive books, or books on other transport or related hobby subjects, are invited to write to Veloce Publishing at the above address.

British Library Cataloguing in Publication Data -
A catalogue record for this book is available from the British Library.

Typesetting (AGaramond), design and page make-up all by Veloce on AppleMac.
Printed and bound in the UK.

Visit Veloce on the Web - www.veloce.co.uk

The
Triumph Tiger Cub
Bible

"A personal history of the
Triumph Terrier and Tiger Cub"

by Mike Estall

Metropolitan Borough of Stockport Libraries		
		CLL
000357855	BRA	
Askews	629.2275	
	EST	

VELOCE PUBLISHING PLC
PUBLISHERS OF FINE AUTOMOTIVE BOOKS

Foreword & Acknowledgements

Foreword

My reason for writing this book is to try and put under one roof as much information as possible about the two Triumph lightweight motor cycles, the T15 Terrier and its successor, the T20 Tiger Cub. Over the years these machines have been described in many magazine articles, and have been the subject of a chapter in a couple of books. In view of their popularity - not only when they were being built but also now in later years - a book devoted to them is long overdue.

This book contains as much history as can be gleaned from all available sources so that the reader can trace the background, development, history, strengths and weaknesses from the earliest days of the Terrier, right through to the end of Tiger Cub production some seventeen years later. Restorers will now - hopefully - have a source of information that will enable them to rebuild their machines to as authentic a condition and specification as possible.

During the writing of this work it has been realised that a great deal of detail is missing. Factory records have been lost or destroyed, employees with first-hand knowledge are no longer available and a huge pool of information in the form of correspondence files, specification books, memoranda, drawings, blueprints and other ephemera have gone, probably thrown out in the period leading up to closure of the Triumph and BSA factories.

This book will answer many questions and is a distillation into one coherent record of all the information that has become available to the author. However, conflicting information abounds, even in extant factory sources, and inevitably some questions will remain unanswered and new ones will be generated. 'Intelligent' guesses have sometimes of necessity had to be made in order to try and set matters straight, but hopefully these deductions will not create further confusion!

Readers with new information or evidence are invited to contact the author via Veloce Publishing so that, one day, perhaps, the full story of the Terrier and the Tiger Cub can be told.

Acknowledgements

When the Meriden factory closed at the end of 1983, most of the records were lost or destroyed. By that time the atmosphere at the factory, which had once been almost magical, had degenerated to the extent that there was virtually a complete lack of care or interest in the product. An army of accountants, multiple-layered management mostly devoid of any motorcycling experience, poor management decisions and powerful trade union interests turned a once-proud name into something of a shambles. At the end any thought of preserving history for posterity had gone and very few records were saved.

The Terrier and Cub had always been regarded as minor models and, by the time the end came, they had been out of production for many years. No thought was given to saving anything that referred specifically to the lightweights so hard factual evidence is now very thin on the ground. With the exception of a few individuals who had been at Meriden for most of their working lives and had retained a strong sense of heritage and pride in their company, there was little interest in salvaging anything from the mess. A very few records did survive and it is from these books and papers, and the memories of some of that small number of people left at Meriden who really cared, that much of the detail in this book has come.

This book would have been much less informative had it not been for the help, advice and information freely given by those ex-employees of the Triumph Engineering Co. Ltd. and BSA Motor Cycles Ltd., and many others too numerous to mention individually. However, there are a few to whom the author is particularly grateful and to whom a debt of thanks is owed, namely: Alastair Cave,

Hughie Hancox, John Nelson, Henry Vale, Jack Wickes and Harry Woolridge.

Grateful thanks are also due to James Hallett in Bermuda for many diligent hours spent researching archive material concerning the legal background to that island's motoring regulations. The author also owes a debt of thanks to Lindsay Brooke in Michigan and Jonathan Brown and Dale Martin in California, who each gave much valuable detail, information and photos of Cub activity in the USA. Thanks are due to Jean Caillou in Paris for all his researches concerning the French Army Cub and for his design of the Tiger Cub and Terrier Register decal, and also to Don Morley and Bert Thorne in Surrey, England for information about the Comerfords Cub. My thanks are also due to John Crichlow, without whom the paintwork details of the various models would have been in a much confused state!

Finally, plaudits must go to my wife, Janet, who spent many patient hours proof reading, listening to ideas, correcting spelling mistakes and errors of syntax and grammar, and spurring me on when enthusiasm waned or I ran out of steam!

M.J. Estall

About the Author

Mike Estall was born in January 1938 in Perivale, Middlesex. He was educated at primary, junior and grammar schools in Greenford, Middlesex, and left school with seven O levels to start work as a laboratory technician in the research laboratories of a paint manufacturer in Perivale. He took physics, chemistry and maths at A level with the intention of going on to do a degree in chemistry but, not being of an academic nature, failed dismally in all these ambitions.

He did two years National Service in the Royal Air Force, serving as an assistant air traffic controller at RAF Tangmere, Sussex. On leaving the RAF in December 1961, he became an officer of HM Customs and Excise, serving in this capacity at London, (Heathrow) Airport, Reading, Southampton, Birmingham and Coventry.

Mike married Janet in 1964 and they have two daughters. He took early retirement in 1991 in order to spend more time with his hobby - old British motorcycles.

Mike's motorcycling activity began on the 19th May 1956 when he bought a brand new Tiger Cub from Pinks of Harrow. Two and a half years and forty thousand miles later, this machine was part-exchanged for a sprung wheel 1953 Triumph 5T Speed Twin, which was kept until Mike's new job in the Civil Service demanded the use of a car. He took little further active interest in motor cycles until one day when, in a moment of weakness, he bought a Tiger Cub 'basket case' to play with. Strangely - and coincidentally - this event took place on 19th May 1981, exactly twenty five years later, to the day, that he had bought the same model brand new!

This was not only the start of a succession of restoration projects, mostly Tiger Cubs or Terriers, but

Mike and Janet Estall. The bike is a 1954 Terrier which the author converted to 125cc. (Courtesy Jim Davies)

also the beginning of a period of collecting and researching technical information on the Triumph post-war lightweights. Mike became the Vintage Motor Cycle Club Tiger Cub and Terrier Marque Specialist in 1989, started the Register of machines as a private project in November 1991, and now runs a worldwide information service for Cub and Terrier owners.

Contents

Foreword 4

Acknowledgements 4

About the Author 5

Chapter 1 The beginnings 9
1.1   Introduction 9
1.2   The opposition 9
1.3   The Triumph 'Look' 9
1.4   E.T.'s first thoughts 9
1.5   Post-war austerity 10
1.6   Initial designs 10
1.7   Publicity 10
1.8   Design and styling 11
1.9   Design features - policy and design criteria .. 11
1.10 Design features - lubrication system 12
1.11 Design features - sludge trap 12
1.12 Design features - timing side main bearing .. 12
1.13 Design features - primary chain tensioner 12
1.14 Design features - cylinder head and barrel ... 13
1.15 Design features - crankcase and oil seals 13
1.16 Design features - flywheel assembly 13
1.17 Design features - transmission, valve gear,
       gear indicator, dimensions 13
1.18 Design features - electrical system 14
1.19 Design features - cycle parts 14
1.20 Prototype wouldn't start 14
1.21 Prototype variation and production
       machines 14
1.22 Electrical system - Wipac and Lucas in
       pre-production machines 15
1.23 Electrical system - Lucas for production
       models. Alternator wire colours 15
1.24 Electrical system - points location 16
1.25 First showing 17
1.26 The competitors 17
1.27 Assembly track 18
1.28 Delay after launch 18
1.29 The 'minimum amount of metal' policy 18

Chapter 2 Methods and procedures 19
2.1   Financial year 19
2.2   Acquisition of the machine serial number
       and model type 19
2.3   Order of building on the track 20
2.4   Frame and engine numbers described 20
2.5   Location of frame and engine numbers 20
2.6   Registration numbers 20
2.7   First production machine 21
2.8   'Year' number sequences 21
2.9   Numbers after 100013 summarized 21
2.10 Numbering and model type anomalies.
       Gaps in number ranges 22
2.11 Different frame and engine numbers 22
2.12 Duplication of machine numbers 22
2.13 Differences in model type 22
2.14 Financial year & machine numbering.
       Clerical errors 23
2.15 Register of machines 23
2.16 New season's colour schemes 23

Chapter 3 Taking the plunge 26
3.1   Plunger suspension 26
3.2   Overall gearing 26
3.3   The Tiger Cub introduced 28
3.4   Tiger Cub - first impressions 28
3.5   Tiger Cub - delay in early sales 28
3.6   Press road tests 28
3.7   Problem areas - general 29
3.8   Problem areas - lubrication and oil leaks...... 30
3.9   Problem areas - excessive oil pressure 30
3.10 Problem areas - 'wet-sumping' 30
3.11 Problem areas - clutch and primary drive 32
3.12 Problem areas - electrics 32
3.13 Problem areas - rear suspension, centre stand,
       frame, cables, rear chain 33
3.14 Problem areas - big ends and flywheel
       assemblies 33

3.15   Problem areas - troubles due to
        performance 34
3.16   Problem areas - noise 35
3.17   Problem areas - other 35
3.18   Problem areas - summary 35
3.19   Total sales of the plunger models 36
3.20   Fochj - 'The Italian job' 36

Chapter 4 The Gaffer's Gallop 39
4.1   The runners and riders 39
4.2   The route 39
4.3   The performance 40
4.4   The results 40

Chapter 5 Full swing at Meriden 43
5.1   Design changes - new swinging fork frame,
        steering lock 43
5.2   Design changes - hydraulic front fork 44
5.3   Design changes - two-piece crankcase.
        Experimental paper oil filter 44
5.4   Design changes - general improvements 45
5.5   Design changes - silencing 45
5.6   Design changes - carburettor 46
5.7   Design changes - 'works' trials machines 47
5.8   Design changes - petrol tank and
        badge type 48
5.9   Design changes - side panels 49
5.10   Design changes - large inlet valve 51
5.11   Design changes - new crankcase and
        'two-ball' mains. Revised oilways 51
5.12   Design changes - new crankcase with
        'side points' 55
5.13   Design changes - twin switches, gear
        indicator move, new alternators 55
5.14   Design changes - cylinder barrel and head
        finning 55
5.15   Design changes - wheel and tyre sizes, brake
        drums and speedometer drive 55
5.16   Design changes - pistons, cams and gearsets 56
5.17   New models - T20 & T20J 57
5.18   New models - T20C and derivatives
        T20CA and T20CB 58
5.19   New models - T20S 61
5.20   New models - T20S derivatives - model
        types and specifications 62
5.21   New model - the origins of the Mountain
        Cub 64
5.22   New models with Energy Transfer systems .. 66
5.23   Energy Transfer - operating principles 66
5.24   Prototype - the two-stroke twin 67
5.25   Prototype - the overhead cam twin 68

Chapter 6 Built at BSA 69
6.1   Move to Small Heath - reasons behind the
        decision 69
6.2   Move to Small Heath - warranties and
        build quality 69
6.3   Machine numbering - different to Triumph 70
6.4   New model - T20B Bantam Cub 70
6.5   New model - T20B Super Cub 70
6.6   Bantam and Super Cubs - their
        undeserved unpopularity 71

6.7   The other Small Heath models 71
6.8   New model? - the 'Tarbuk' Conversion 72
6.9   New model? - T20 Bantam 175 74
6.10   Prototype - 'Pastoral Cub' 74
6.11   Total sales from Small Heath 74

Chapter 7 Stateside 75
7.1   The American market 75
7.2   USA machine specification 75
7.3   Triumph Corporation - 'TriCor' 75
7.4   Johnson Motors - 'JoMo' 76
7.5   Sales to the USA 76
7.6   Early machines rationed 77
7.7   Export packing 77
7.8   Announcement of the Tiger Cub in
        the USA 77
7.9   Dedicated USA models 77
7.10   Colour schemes 78
7.11   After-market performance parts 78
7.12   Competition in the States 80
7.13   Types of events in the USA 81
7.14   Major Enduro successes 81
7.15   Cubs on 'Short-tracks' 82
7.16   Other events 84
7.17   A world speed record 85

Chapter 8 The 'Bermuda Cub' 89
8.1   Capacity and bore limits 89
8.2   Special model, numbers delivered 89
8.3   Legislation from 1905 to 1947 89
8.4   1946 legislation, definitions 90
8.5   First demonstration. 1946 legislation,
        speed limit 90
8.6   1949 legislation, pedals, engine power .. 91
8.7   1951 and 1953 legislation, no pedals,
        weight limits 91
8.8   1956 and 1973 legislation, capacity
        limited to 150cc 91
8.9   Noise limit 91
8.10   Legislation - summary 91
8.11   The first imports by Charles Young 91
8.12   Engine configuration - '150' and engine
        types 92
8.13   Engine configuration - barrel, pushrods
        and flywheel assembly 92
8.14   Cycle parts 92
8.15   Sales - general 92
8.16   Bermuda today 92

Chapter 9 The Cub in competition 94
9.1   General 94
9.2   1953 96
9.3   1954 97
9.4   1955 98
9.5   1956 99
9.6   1957 100
9.7   1958 100
9.8   1959 101
9.9   Late-fifties and early sixties 102
9.10   1960 102
9.11   1961 102
9.12   1962/3 104

9.13   1964/5 104
9.14   1966 onwards 104
9.15   The 'Comerfords Cub' 105

Chapter 10 Military, Police, and the Utilities..... 107
10.1   The French Army Cub 107
10.2   Orders and first deliveries 108
10.3   Technical details 108
10.4   Colour scheme 108
10.5   Machine identification 108
10.6   Machine maintenance 109
10.8   Problems with electrics 109
10.9   Problems with oil leaks 109
10.10   Problems with centre stands 110
10.11   Problems with overheating...................... 110
10.12   Cessation of contract 110
10.13   The French Army Cub delivery total 111
10.14   The French Army Super Cub? 111
10.15   The French (dis)Connection 111
10.16   The French Army Cub today 111
10.17   Cubs in the British Army 111
10.18   T20WD, features used on later models 112
10.19   T20WD, testing 112
10.20   T20WD, purchase by other armed forces .. 113
10.21   Cubs in the Constabulary 113
10.22   Police Terriers 114
10.23   Police Cubs in Cambridge 114
10.24   The RUC 114
10.25   The T20P 114
10.26   T20P livery and equipment 114
10.27   Police Super and Mountain Cubs 115
10.28   Police machines - summary 115
10.29   Local Authority users 118
10.30   Summary.. 118

Chapter 11 In memoriam 119
11.1   Early hopes 119
11.2   Trouble on the horizon 119
11.3   The new Japanese industry 119
11.4   Sales to the end 120
11.5   Cub deliveries abroad............................. 120
11.6   The end of the line 120
11.7   The Tiger Cub today 122

Appendix 1   Model profiles 125
App.1.1   T15 1954 to 1956 125
App.1.2   T20 1954 and 1955 127
App.1.3   T20 1956 129
App.1.4   T20 1957 131
App.1.5   T20, T20J 1958 133
App.1.6   T20, T20J 1959 135
App.1.7   T20, T20J 1960 137
App.1.8   T20 1961 and 1962
           T20J 1961 139
App.1.9   T20 1963 onwards 141
App.1.10   T20C 1957 to 1959 143
App.1.11   T20CA 1958 and 1959 145
App.1.12   T20S 1959 to 1961
           T20W 1960 147
App.1.13   T20T 1961 149
App.1.14   T20SL 1961 151
App.1.15   T20SS 1962 to 1965 153
App.1.16   T20SH 1962 to 1966 155
App.1.17   T20SR, T20SC 1962 to 1965 157
App.1.18   TR20, TS20 1962 to 1965 159
App.1.19   T20SM, T20M 1964 to 1967 161
App.1.20   T20M.WD Fr. Army 1964
           to 1967 163
App.1.21   T20B Bantam Cub 1966 to 1968 .. 165
App.1.22   T20B Super Cub 1967 to 1969...... 167

Appendix 2   Production year - first and last serial
              numbers. Model types 168
Appendix 3   Component dimensions 170
Appendix 4   Carburettor - types and settings ... 177
Appendix 5   Model colour schemes................. 178
Appendix 6   Delivery details - by country
              of delivery 180
Appendix 7   Delivery details - by number of
              destination countries................... 193
Appendix 8   Published road tests 194
Appendix 9   Chronology of specification
              changes.................................... 195
Appendix 10   Useful addresses 205
Appendix 11   Further reading 205

Index 206

Visit Veloce on the Web - www.veloce.co.uk

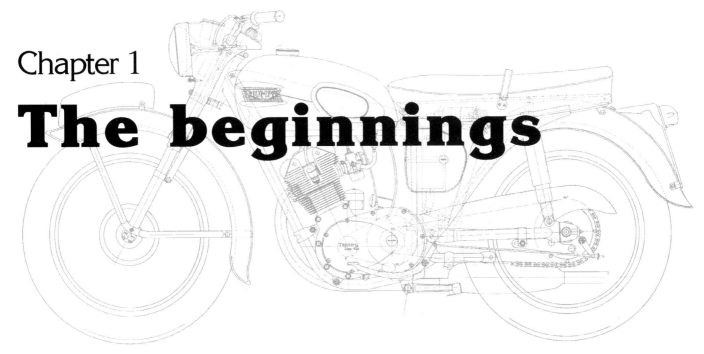

Chapter 1
The beginnings

1.1 Introduction

Every now and then there appears a motorcycle destined for stardom; a machine which, when compared to the others in its class, stands head and shoulders above them. One such machine was the Triumph Terrier and, to an even greater extent, its successor, the Tiger Cub.
When viewed in the light of history, it is obvious that the Terrier was destined to be a winner and a trend-setter. Not that it was perfect: like all machines it had failings, many of which were apparent from the start and should have been corrected before the new model went into production. But when prospective new owners compared the Terrier or the Cub to the opposition, there was little room for doubt.

The Terrier's basic engine design continued for years in various enlarged and modified forms, spawning a whole range of motorcycles built by BSA, Triumph's parent company, and others. For example, the engine units of the C15, B25, TR25W, B40, B44, B50, CCM, etc., all derived directly from Edward Turner's original Terrier design, the 250cc C15 being the first to arrive in 1957, replacing the almost antedeluvian C10 and C11 models. It is worth noting that these engines were developed from the initial 150cc of 1952 up to a full 499cc by the end in 1974.

1.2 The opposition

Consider what machines were available in late 1952 in the 125 to 150cc capacity class. Without exception they were two-strokes, mostly powered by the ubiquitous Villiers engines. Of these machines only one, the BSA Bantam, could be described as having been really successful, and only then by using an engine whose original design had been borrowed from another European maker. The 149cc T15 Terrier was the only British lightweight four-stroke among a plethora of smoking two-strokes. It also fitted very nicely inside the cheap (17/6d (87.5p)) 150cc road tax bracket - an important factor in times when low running costs were paramount.

A year after introduction of the Terrier came an uprated or sports version, the 199cc T20 Tiger Cub, appearing at the November 1953 Earls Court Show. This time it was up against slightly stiffer opposition - there were other four-strokes to be taken into account - but many potential purchasers held the opinion that none could match the Cub when it came to setting cost against performance. But more of this model later.

1.3 The Triumph 'Look'

With their brightly coloured paintwork, light weight, ease of handling and extremely lively performance for the day, coupled with very good economy, the Cub and Terrier were much desired. On top of that they had the 'look' of a Triumph. The 500 and 650cc twins from this manufacturer were considerd by many to be the finest looking machines in their class. They had style, together with a cleanliness of line, and the new lightweights came straight out of the same mould.

To many young riders of the day, (including the author), these little motorcycles, with their good looks and performance, were just about irresistible! Riding one transported the owner into the charismatic world of Triumph motor cycles. So, how did it all come about?

1.4 E.T.'s first thoughts

It's not known precisely when thoughts of a new lightweight model first entered Edward Turner's mind. There had not been a small machine in the Triumph range since the 250cc pre-war models and the Model XO 150cc single cylinder four-stroke of the same era. Turner was an astute judge of the market and, in the summer of 1952, the climate must have seemed just right to him.

Fig 1.5 The Meriden factory, built 1942. (The Mike Estall Collection)

As long ago as 1943, in a paper to the Institution of Automobile Engineers, Turner had postulated that a 'Primary motor cycle up to 150cc capacity', would be one among five categories of machine that could be made to cover the entire post-war market spectrum. He had obviously been thinking about a new lightweight 'safety' model for some time, and in mid-1952 the decision was taken to go ahead. The design team had Edward Turner at the helm, Harry Summers as chief draughtsman and Jack Wickes as the styling engineer. Things then began to move very quickly indeed!

1.5 Post-war austerity

England was just emerging from the severe depression that had followed the end of the war. Raw materials and the resources necessary to make new motor cycles were in short supply, but people needed transport and new cars were both scarce and expensive. It was, 'Export or Die', and every new vehicle made here was being sent abroad to earn precious foreign currency.

It was much the same in many other parts of the world. Only a few western countries had the capacity to build cars and motor cycles in quantity and the United States, with the largest single market of all, only made thundering great V-twin motor cycles and huge, fuel-gobbling cars. Turner judged that there was clearly an opening for a small commuter motor cycle with sufficient performance to easily carry two people yet not consume great quantities of fuel.

1.6 Initial designs

It was in this atmosphere that Edward Turner made the first thumbnail sketches and laid out the design principles of his new model, the 150cc Terrier, in the summer of 1952. The machine was designed as a commuter model

with the emphasis on reliability, safety and economy. Contemporary sources reported the actual time from first sketches to running of the prototype as being only eight weeks, an incredibly short period!

It is possible that this very short time was achieved because Triumph may have put pressure on its French importer, Peugeot, which was in a position to persuade another French company, Terrot, to supply drawings of a Terrot design, which was, perhaps, similar to that of the Terrier. However, it must be said that this is pure speculation and may be completely erroneous. We shall probably never know the new model's true origins ...

In those eight weeks hundreds of engineering drawings and subsequent revisions would have been produced, styling decisions made, wooden patterns constructed for the engine castings and other items, and all manner of other engineering tools - dies, gauges, jigs and other parts - made both in-house and by sub-contractors. Then the whole lot would have been brought together by the factory's Experimental Department some time in September or October 1952.

1.7 Publicity

In the meantime the Publicity Department had to get cracking, distributing information about the new machine to the motor cycling press. Word leaked out in late October but the official announcement was made in the first week of November 1952, just prior to the Earls Court Show.

Articles appeared in *Motor Cycle* and *Motorcycling* magazines giving full technical details with photographs and drawings. Sales catalogues and posters, which took around three months prepare, had to be produced and none of this could be done until all the design details had finally been settled. It must have been a busy time all round!

Fig 1.7a A pre-production Terrier without a gear indicator. Taken from the first sales brochure - December 1952. (The Mike Estall Collection)

1.8 Design and styling

There were various design and styling exercises, mainly done by Jack Wickes, Harry Summers, Frank Thompson and freelance artist Roy Coombs, including one with heavily flared front and rear mudguards. The final design followed the Triumph 'look' having, like the larger twin cylinder models from October 1948, the streamlined instrument nacelle. It was a scaled-down version carrying the headlamp unit, switchgear and speedometer, (and later, on production machines, the gear indicator). It gave most of the Triumph models a very 'clean' cockpit area and hid much of the untidy cable clutter visible on so many other makes.

There was a single saddle giving a comfortably low seat height, the Triumph 'Four-Bar' petrol tank badges, discreet use of chrome plating and an Amaranth Red paint job, just like its larger sister the 5T Speed Twin. It also had such ground-breaking features in a lightweight as an overhead valve four-stroke unit construction engine and gearbox with dry sump lubrication, four speeds and a crankshaft mounted alternator. It was a very good-looking machine, strongly reminiscent of the rest of the Triumph range.

1.9 Design features - policy and design criteria

The heart of the Terrier was the new engine and gearbox unit made in die-cast aluminium. It was Triumph's first venture into unit construction since before the second world war, and the new engine was remarkably small and

compact with some rather neat design features.

Turner had decided against a two-stroke on the grounds of performance versus capacity. Although a two-stroke would be cheaper to produce, he considered that the benefits of the overhead valve four-stroke (better performance, lower running costs, positive lubrication and ease of silencing), outweighed the initially higher production cost. He also preferred, 'bearings to be lubricated by green oil instead of lightly coloured petrol'. The side-valve configuration was dismissed because of its asymmetric cylinder section which might cause distortion problems when working hard at large throttle openings.

Unit construction and the nearly 'square' bore and

Fig 1.8 An artist's impression of a pre-production Terrier. Note the Wipac electrics, flared front mudguard and no gear indicator. (The Mike Estall Collection)

stroke dimensions were chosen to keep the overall dimensions small and to give a smooth and easily cleaned engine/gearbox outline. They also came about from a desire to produce the smallest engine possible for a given power and to provide the greatest area for the dissipation of excess engine heat.

To sum up, Turner considered that unit construction was stiffer, lighter, cheaper and cleaner. He may have been right having faith in the last three of these proclaimed characteristics but events were to prove that he got the sums wrong when it came to the first.

1.10 Design features - lubrication system

The engine featured dry sump lubrication. A double plunger type oil pump pulled oil from an external tank, forcing it through the bushed timing side main bearing, via drillings into the big end. Oil would then splash onto the lower bore area and the drive side main bearing before draining down to be scavenged back to the oil tank. The overhead valve gear was lubricated by a feed taken from the oil tank return line. A skew gear, machined as part of the crankshaft pinion, drove both the oil pump and an ignition contact breaker mounted on top of the crankcase.

1.11 Design features - sludge trap

A sludge trap in the timing side flywheel was not specified until April 1954 at engine no. 4859. Big ends had been seizing with some regularity and one possible cause could have been foreign matter carried in the oil.

At this point Edward Turner applied an idea no doubt picked up during his spell at Ariel with Val Page. The sludge trap, which had been a feature of all Ariel single cylinder flywheel assemblies since 1933, removed particles of foreign matter in the circulating oil by using the rotating timing side flywheel as a centrifuge, thus drawing any muck out of the oil and into the chamber. It

was recommended that the chamber, which was closed off by a screwed plug whose head was centre-punched for security, should be opened periodically to allow the accumulated sludge to be removed. This was an onerous but necessary task as detergent oils were not yet in everyday use, and waste combustion products could do nothing else but accumulate in the oil to be recirculated around the engine, quietly grinding away all the bearing surfaces. Naturally, this was seldom done as a matter of routine but performed only occasionally and when breakdown maintenance was required, by which time the damage had already been done ...

1.12 Design features - timing side main bearing

Even before the first metal was cut there was one design feature which, had it continued into production, might have led to disaster. Edward Turner had designed the timing side of the crankshaft to run directly in the metal of the crankcase, without the benefit of a separate timing side main bearing! Some stationary engines ran without one but a motor cycle engine in 1952 was something altogether different. Fortunately, this shortcoming was corrected in time and a steel-backed white metal bush was inserted at this point.

Had this feature not been included the consequence would have been that, as the flywheel assembly rotated, the crankcase itself would have worn, requiring replacement of the complete engine casing to restore oil pressure to the big end. Integrity of fit between the flywheel stub and its housing was crucial to the maintenance of good oil pressure and inclusion of a replaceable bush was indeed wise. When, in later years, unit construction was extended to the twin cylinder 'C' range, a replaceable bush was included as a matter of course.

1.13 Design features - primary chain tensioner

Features on the inner drive side cover for the one-piece crankcase indicate that a primary chain tensioner had been planned. This cover had on its outer side two small platforms situated immediately above a thickening of the lower edge. Clearly, a slipper-type primary chain tensioner had been considered at some early stage, sliding on the two platforms and being screwed up and down by an adjuster through the thickened lower part of the cover.

This would have been a useful addition in the early days because primary chains tended to 'grow' in use. Wear between the rollers and pins and stretching of the side plates made the chain slack: in extreme cases a worn chain could actually strike the chaincase. To some extent the problem was alleviated in 1956 and 1957 by the use of a pre-stretched chain, and later by increasing chain clearance of the inner and outer drive side covers. But the simplex chains were not really up to the job, even when uprated from $^3/_8$in x $^7/_{32}$in to $^1/_2$in x $^3/_{16}$in. The arrival of the duplex chain did much to improve the situation, but if lubrication was not properly attended to this chain also wore and stretched.

Fig 1.13 T15 engine arrangement. Note the adaptations in the primary side inner cover for a chain tensioner. (The Mike Estall Collection)

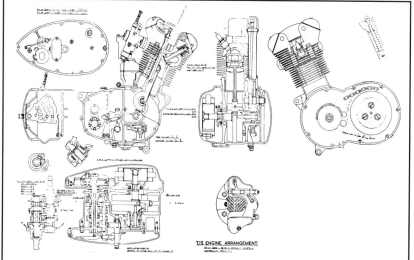

1.14 Design features - cylinder head and barrel

A cylinder head with integral rocker boxes (like the 3T and 3HW models before it), in die-cast aluminium sat atop a silver painted cast iron barrel which sloped eagerly forward at 25° from the vertical and was roughly parallel to the frame front down tube. The cylinder head was finned on the exhaust rocker casing but not on the inlet side, and there were chevron-shaped fins immediately over the combustion hemisphere to assist with cooling.

Valve elevation generated by the camshaft was further enhanced because the rocker arms were asymmetric; one side being longer than the other. The pushrod movement from the cam lobe was increased by a further ten percent due to the extra length of the arm on the valve side of the rocker.

The cylinder barrel was the 'round' shape and the barrel studs were visible through the cooling fins. This barrel continued until the 1959 season when the 'oval' type came into use, itself to be superceded by the 'square' type in early 1965.

1.15 Design features - crankcase and oil seals

The one-piece crankcase design was very clever. It was constructed in unit with the gearbox and had a removable inner cover on the drive side which carried the left main crankshaft bearing. This cover was spigoted into the crankcase with a 0.0027in interference fit and was secured by screws.

The design was such that, once removal of the left side outer cover, alternator rotor, clutch and primary drive, right side inner and outer covers with gearbox contents, cylinder head, barrel and piston had been accomplished, the flywheel assembly could be withdrawn. The whole engine could be taken apart without having to remove it from the frame: very neat.

Initially, an oil seal had not been fitted to the drive side engine main bearing, although one was very soon provided as it was found that the opening period of the timed breather in the crankcase could not deal with all the pressure variations caused by the piston pumping action. Crankcase pressure could fall below that found in the primary chaincase, gradually drawing oil out of the chaincase and into the crankcase through the main bearing and mixing the chaincase oil with the engine oil being scavenged back to the oil tank. Fitting an oil seal at this point thus prevented the primary chaincase being slowly drained of oil.

There was no oil seal behind the clutch until February 1955 when a rubber disc and steel washer came into use. This arrangement did not last very long, being replaced in August 1957 by a rubber 'labyrinth' type seal working on a longer high gear bush that extended through it and into the primary chaincase.

The high gear oil seal was at first made of felt, with a steel backing washer. An attempt had been made in August 1954 to cure leakages here by specifying a thicker felt seal, but it wasn't until August 1957 that a proper garter seal was fitted. Even then the solution was not completely satisfactory and two thinner seals were specified, to be fitted back-to-back. Interestingly, in late 1964, the single seal was brought back into use again.

1.16 Design features - flywheel assembly

The flywheel assembly was a pressed-up affair with the timing side main bearing consisting of an integral stub on the flywheel running in a removable bush pressed into the crankcase. The drive side mainshaft was initially secured in the flywheel by a taper with keyway and locknut. This method of fixing was changed in April 1954 to a serrated flange on the end of the mainshaft which locked into a recess machined into the inner surface of the flywheel. A 1$\frac{1}{2}$in diameter single row uncaged roller big end ran in a forged steel connecting rod fitted with a phosphor bronze small end bush.

1.17 Design features - transmission, valve gear, gear indicator, dimensions

There were four speeds in the gearbox and a clutch with two cork-faced drive plates and two driven plates, plus a rubber vane-type shock absorber. There were pushrod operated overhead valves in the cylinder head and a gear indicator to show the novice rider what part of the box was working. The bore was 57mm and the stroke 58.5mm, giving 149.5cc. The compression ratio of 7:1 was quite high in view of the low octane petrol then generally available.

The whole engine/gearbox unit, which weighed only about twenty two kilograms or just under fifty pounds, produced eight brake horse power and could be taken out of the frame by simply removing the carburettor, exhaust system, rear chain, oil pipes and three engine bolts. It was powerful enough to propel the machine, rider and pillion passenger at a happy fifty mile per hour cruise and gave a top speed of about sixty.

Fig 1.17 The gearchange mechanism. (Courtesy EMAP)

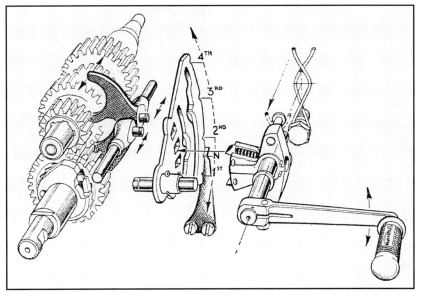

1.18 Design features - electrical system

Electrical power came by way of an alternator supplying AC current to a six volt DC rectified system with a battery. The sparks came from a six volt HT coil via a contact breaker with an automatic advance and retard mechanism. An emergency ignition system allowed the engine to be started even if the battery was flat. This electrical layout, which was lighter and physically smaller than the faithful old magneto, had been pioneered on the 3TW model in 1940. Its only other use had been on the side valve TRW twin just after the war, but this model had still retained magneto ignition. It was later adopted for several Triumph twin cylinder models.

1.19 Design features - cycle parts

The cycle parts into which this little engine was placed bore a strong similarity in style to the big twins. The frame had a single down tube with a pronounced swan neck at the steering head and, unusually for Meriden, plunger suspension at the rear. The Terrier, and the first three years of the Cub, were the only Triumph models fitted with this type of springing. The system worked reasonably well provided it was greased regularly, but inevitably it succumbed in the general trend towards swinging fork frames and by 1957 was gone.

The wheels were 19in diameter with Dunlop block pattern tyres and 5 1/2in diameter brakes in single sided hubs. The front forks were undamped, grease filled and

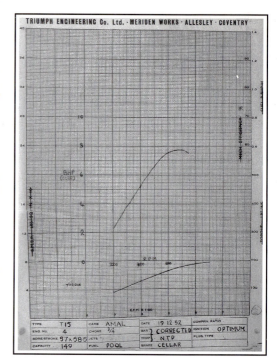

Fig 1.21 T15 torque and power curves from December 1952. (The Mike Estall Collection)

of rather skinny appearance, but quite adequate for the purpose, and the whole package weighed-in at one hundred and seventy five pounds. It had a power to weight ratio which was about the same as the 350cc B31 BSA or Mk.V Douglas and much better than the 150cc BSA Bantam D3 or Royal Enfield Ensign.

1.20 Prototype wouldn't start

Apparently, the first prototype machine refused to start. It was kicked over and pushed up and down for several hours by the men from the Experimental Department and other factory personnel, in an attempt to bump-start it, but not until the following day did it finally spring into life. The reason for this initial failure and how it was cured is not known, but it's most likely to have been electrical in nature - an apparently recurring problem with the pre-production machines. It was later found that the HT coil produced only about 8Kv at the spark plug instead of the 10Kv or more needed, and this may have been a contributory factor.

1.21 Prototype variation and production machines

There were several differences between the prototypes and later machines as they appeared in production. The pre-production machines tried both Wipac and Lucas electrics, some had a non-folding kickstart lever and all were devoid of a gear indicator. Some of the frame lugs were heavy pressings which, in production, became

Fig 1.20 Jack Wickes on the Terrier prototype. Photographed around October 1952 on the back lawn of the Meriden factory. (The Mike Estall Collection)

castings. On the pre-production machines the rocker covers were secured by knurled knobs straight off the 3T twin, and the rear suspension covers were polished alloy. Initially, a Lucas MT110 rear light was used, later substituted by a 525 or 529 type. The ignition coil was at first mounted under the saddle whereas early production machines carried it on the rear mudguard behind the gearbox.

The illustration on the front cover of Parts Lists no. 1 and no. 2, dated October 1953 and 1954, shows a tube, perhaps a breather or oil feed pipe, in place on the cylinder head above the pushrod tunnel. This feature did not appear in production and its function is uncertain.

An eighteen tooth gearbox sprocket was initially specified, giving a top gear of 6.7:1 and 900 engine rpm per 10mph road speed. This was changed in early production machines to seventeen teeth, giving a 7.1:1 top gear and 953rpm per 10mph. The power curve showed 7.92bhp at 6500rpm on 'Pool', (approx. 80 octane), petrol. The torque curve was fairly flat showing 90lb/in at 3000rpm, rising to 165lb/in at 5000rpm and 185lb/in at 6500rpm.

A gear indicator did not appear on the pre-production machines but in May 1953 it was specified for use on all production Terriers. This device, which the experienced Triumph rider might have considered something of a gimmick, was very useful to the novice. It worked by a control rod which was attached to the gearbox camplate, transmitting the camplate motion up to the instrument nacelle by means of a cable. Here, the movement of the cable inner was converted into circular motion by a rack and pinion mechanism, which in turn drove a pointer around a dial marked with the gear positions. If properly adjusted the indicator would show which gear was engaged, if not it would hover meaninglessly between gears. Worse still, if the cable had been adjusted too tightly, top gear could not be engaged as tension in the cable would prevent the camplate from achieving its full motion.

All nacelle models were fitted with this type of indicator, right from T15 no. 101 to T20 no. 88346 at the end of the 1962 season. For 1963 and 1964 the indicator became a rising and falling indexed finger poking out from the top of the gearbox and really was not much use at all to anyone. By 1965 it had disappeared altogether.

1.22 Electrical system - Wipac and Lucas in pre-production machines

Six pre-production Terriers were built; the first three with Wipac electrics and then three with Lucas. The initial advertising campaign of late 1952 had loudly declared that the new model would be fitted with a Wipac electrical system but in production this was not to be.

The Wipac series 110 AC generator fitted gave, at 2000rpm, 54 watts at 6 volts rising to 58 watts at higher speeds. The alternator lead on Wipac machines exited the front of the primary chaincase cover just below its centre line, whereas the Lucas alternator lead left the top of the case at the back of the drive side inner cover. The Wipac HT coil gave 9000 volts in the emergency start position at kickover speeds. The battery was to have been a Varley MC5/9.

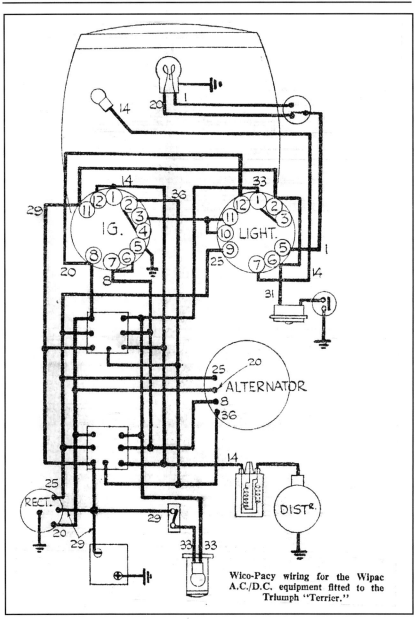

1.23 Electrical system - Lucas for production models. Alternator wire colours

Evaluation testing showed that Lucas equipment was superior in sufficient aspects to become the preferred fitting on all production machines. The only exceptions were to be some T20B Bantam and Super Cubs, which were fitted with Wipac components from April 1967.

The Lucas RM13, (Rotating Magnet), alternator provided alternating current, which was then converted by a rectifier to direct current supplying the six volt battery system. The battery was a type Lucas PU35E, later a PU5E. The PRS8 combined lighting and ignition switch was used as standard right up to the end of July 1962, after which came two 88SA type switches for 1963

Fig 1.22 A wiring diagram for the Wipac system that did not go into production. (The Mike Estall Collection)

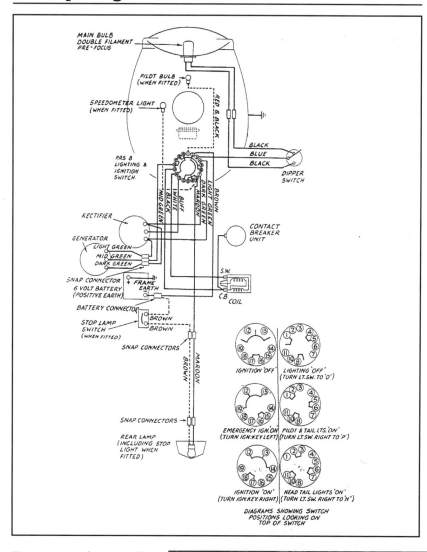

and subsequent seasons.

Over the years the wire colouring of the three-wire Lucas alternator changed. Initially, there had been three shades of green - light, medium and dark, but it was very soon found that under the influence of heat and oil these three colours became one - murky green! So to make things clearer the medium green wire was given a yellow tracer, becoming green/yellow. Later still the other wires gained tracers too. In the late sixties Wipac equipment was again specified and wire colours were different. The complete pattern was as shown in the accompanying table.

1.24 Electrical system - points location

Wire	1953 Lucas	Aug 1955 Lucas	Approx. 1963 Lucas	Oct 1966 Wipac
1.	Light Green	Light Green	Green/White or White/Green	Green
2.	Med. Green	Green/Yellow	Green/Yellow	Orange
3.	Dark Green	Dark Green	Green/Black	Yellow

The contact breaker was located in a mushroom-shaped housing on top of the crankcase. It was driven by a skew gear machined as part of a timing pinion bolted onto the end of the crankshaft. The housing could be easily rotated to adjust the ignition timing and was locked in position by means of an external clamp.

Although this position had advantages in terms of ease of maintenance and adjustment (also preventing any risk of oil seepage onto the electrical contacts), it did spoil the engine's cleanliness of line. Turner had gone to considerable trouble and expense to put this item here, perched on top of the crankcase, explaining that it was done on the grounds of accessibility and keeping the overall width of the engine unit to a minimum. Later experience, particularly with the precise timing require-

Fig 1.23 An early wiring diagram for the Lucas PRS8 switch, which was modified several times up to the end of the 1962 season. (The Mike Estall Collection)

Fig 1.25a A *Triumph News* (Meriden's own newspaper), feature from Earls Court. (The Mike Estall Collection)

ments of the Energy Transfer system, was to show that the later location of the contact breaker on the end of the camshaft would have been a far better choice from the beginning.

1.25 First showing

The Terrier was revealed to the world at the November 1952 Earls Court Show. This show was the largest of its type in England and was by custom the place at which motor cycle manufacturers first showed their new models.

The machine exhibited was not the initial prototype but a highly polished 'amended' prototype, one of the Wipac equipped variants. The new model took the show by storm: interest was such that buyers had difficulty getting close enough to examine the new product in detail. Later in the show a second machine was brought in and placed in the company's show office, spreading the load and making it easier for potential export buyers to examine the new model more closely.

None of the necessary production infra-structure had yet been set up at Meriden so deliveries were not promised until the spring of 1953. Clearly, with such a long time between the show and first anticipated deliveries, Triumph was fully aware that the new model on display was still in its infancy and that much development work would be required before it would be ready for marketing.

1.26 The competitors

The price of the Terrier had already been declared as £98 plus Purchase Tax, totalling £125.4.6, (£125.22 in modern parlance), with a provisional price to the USA of £73.10.0 (£73.50), plus packing. By 1954 taxation changes had reduced the Terrier price to just over £117 and the Cub to just over £127, comparing quite favourably with the opposition.

Fig 1.25b The new Terrier on display at the Dutch show. (Courtesy EMAP)

For example, the Ambassador 197cc Supreme was nearly £149, the Francis-Barnet 197cc Falcon 67 was £124 and a 197cc James Commando was £138. However, you could also buy a 148cc BSA Bantam D3 or a Royal Enfield Ensign 148cc for just over £94 and a Tandon Imp Supreme 197cc for just under £104. But these were all two-strokes.

The range of two-strokes in competition with the Cub was even greater than had been lined up against the Terrier. Manufacturers such as James and Francis Barnet, Excelsior, Bond, Bown, Norman, Tandon and others, produced machines in the 200 to 250cc capacity range.

In 1954 the only British 200cc four-stroke models available were the Ariel Colt at £132 and the LE Velocette at £163. There were no 150cc four-strokes. However, the Cub did have to contend with several 250cc four-strokes such as the side valve C10L BSA at £114 or the overhead valve C11G model at just under £124, also the plunger framed side valve Indian Brave at £139. On offer, too, were the OEC Apollo at £142, the Panther 65 for nearly £136, the Royal Enfield plunger framed Clipper at £162 and the overhead cam NSU Max, marketed by Vincent at a massive £228. With machines and prices like that, and the Triumph image, style and performance, it's no wonder that both Meriden lightweights were in such demand!

Profits on the sale of each Terrier or Cub were somewhat less than those made on the Triumph twin cylinder machines and the lightweight models have therefore to be considered 'loss-leaders'. However, they created new Triumph owners, generating marque loyalty and the possibility of future sales of larger machines to the new rider.

1.27 Assembly track

A new and separate assembly track had to be established at the Meriden works for the lightweight models, which had become the Triumph 'A' Range of motor cycles. The track was situated roughly in the centre of the factory and adjacent to the Paint and Service Repair Shops. By the time production got into full swing this track would be capable of turning out up to two hundred and fifty machines a week. It was in use right through to the transfer of Cub production to BSA's Small Heath works in February 1965.

1.28 Delay after launch

Despite the long delay between the launch of the new model and the first sales (a gestation period of about nine months), many problems remained unsolved, infecting the production machines. Certainly there would have been extensive testing and evaluation of the new model and a continual series of modifications to correct any weaknesses found, but it all took time.

The fabric of the factory itself would have been altered to some extent to accommodate a new assembly track. Stores requisitioning and assembly procedures had to be set up too, all of which would no doubt have had their own teething troubles. The factory was at the same time very busy with its line of large capacity twins, so perhaps insufficient resources were available to devote to development of the new T15 model. Whatever the reasons for this long delay, many initial modifications and even some later ones were not always satisfactory in effecting a cure for the machine's ills.

1.29 The 'minimum amount of metal' policy

From his very early days in 1927 as Jack Sangster's chief development engineer at Ariel Edward Turner had maintained the principle of saving costs by paring metal, making the 'minimum amount of metal perform the maximum amount of duty'. This was all very laudable, enabling an increased profit to be made on each motor cycle built, but it was stretched too far on this occasion and the result was a seriously under-engineered machine.

This axiom caused the Terrier to come into this world scarce half made-up and with many inbuilt faults that should have been eliminated long before any sales were made to an eager but unsuspecting public. As will be seen, there were many later problems that could be laid directly or indirectly at the door of Edward Turner's philosophy.

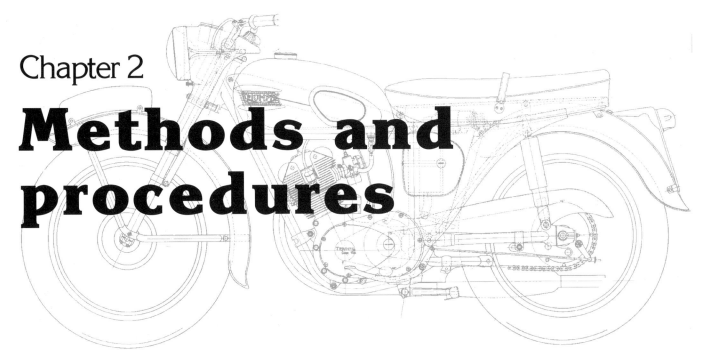

Chapter 2
Methods and procedures

2.1 Financial year

The financial year for all motor cycle manufacturing businesses within the BSA group ran from 1st August to 31st July. This was the basic accounting period for all the group's activities, in both monetary and physical terms, and the annual model 'seasons' for the factories was this same twelve month period.

2.2 Acquisition of the machine serial number and model type

1. Orders received from the main agents were entered into an order book.
2. Orders were collected until there were sufficient for a production run of one particular model.
3. When the length of the production run had been decided the model types and serial numbers were allocated, entered into the order book and a build book, and machine tallies made out. The tally showed the model type, basic specification details including colour, customer's name and order number.
4. Job cards were made out listing any manufacture of parts for assembly into the new machines and the requisitioning of existing parts from the stores.
5. Machine serial numbers, as well as being entered in the order book, were put into the despatch book. This record would later be completed with model type, invoice number, despatch date and customer's name. (Most of the surviving records are despatch books).
6. Engines were built on the engine track and the serial number/model type details were taken from the build book and stamped onto the crankcases.
7. Frames and cycle parts were assembled on their track, the engines installed into the frames and the serial numbers taken from the engine and stamped onto the frame before the petrol tank was fitted. In this way the machines came

Fig 2.3a The young worker on the Cub engine track is believed to be 'Pip' Ellison. The photo was taken on September 24th 1957 by the author on a visit to Meriden with the West Middlesex branch of the Triumph Owners Club. (The Mike Estall Collection)

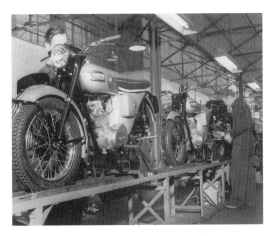

Fig 2.3b 1959 Home market T20 machines coming off the Meriden assembly track. (Courtesy John Nelson)

off the assembly track with matching frame and engine numbers.

8. When testing and final adjustments had been made the machine's tally was attached for identification of the machine in the storage area, before packing and despatch to the customer.

9. When payment had been received the machine was identified from the tally and sent to the packing department for crating, wrapping or preparation for despatch.

It should be noted that Triumph's policy regarding spare new or repaired frames and new spare crankcases was to issue them un-numbered, but to accredited dealers only. The dealer then had to ensure that when sold these items were stamped with the appropriate machine number so as to avoid trouble with the police or Ministry of Transport.

2.3 Order of building on the track

Machines on the assembly track were lined up in model sequence and - once the batches of machines had been selected for building - this sequence of model types followed a set pattern: roadsters were followed by sports models and then it was back to roadsters. For example, a typical building sequence might be T20, T20SS, T20SC, T20SR, T20SH, TR20, TS20, T20SR, T20SS and back to T20 again.

2.4 Frame and engine numbers described

At Meriden the first production machine to start the number sequence was T15 101, in July 1953. Thereafter a single continuous series of numbers was used up to number 100000 (actually 100013), in October 1964, the various model types being allocated within this single sequence of numbers.

The frame number started with the letter "T" and was followed by the serial number: *e.g.* T 12345. However, at around number 95000 in the early part of the 1964 season the BSA parent company's method came into use and the model type then became a prefix to the frame number: *e.g.* T20 98765. All machines built at Small Heath had the model type as part of both the frame and engine number.

Throughout production the engine number at both factories always included the model type; for example, T15 12345 or T20 98765.

2.5 Location of frame and engine numbers

The frame number for the all plunger machines and for the Cub-framed swinging fork models was located on the left hand side of the headlug casting, adjacent to the F3380 or F4421 headlug casting number. On the Bantam-framed machines it was stamped vertically on the left side of the front engine mounting.

The engine number at Meriden was initially located on the top surface of the front engine mounting lug. Very shortly after the earliest machines were built it was

Fig 2.5a Checking the tappet clearances on a Bantam Cub engine. The frame number T20B 245 is stamped vertically on the front engine mount. Part of the engine number is just visible on the crankcase side, just below the bottom fin. (The Mike Estall Collection)

moved to the left hand side of the crankcase just below the barrel, only to be moved back again to the top of the front engine mounting after about the first year of production. There it stayed until manufacture moved to Small Heath when it once again relocated to its former position on the left hand side of the crankcase, below the cylinder barrel.

2.6 Registration numbers

In a very few instances the Triumph and BSA despatch records show a registration number, where a machine was actually registered for the road by the factory. These might be, for example, press road test machines, those sold to employees or machines used by the factory itself. Altogether there are forty six such registrations shown in the despatch books. In all other cases the registration number would have been obtained by the local dealer at the time of sale to the final customer.

At least one instance has been found of what could be described as an example of good tax management. Here, a plunger framed machine was reframed and used as a test bed to try out some parts in a swinging fork chassis. In this case the old frame number was stamped on the new frame, thus avoiding the creation of what HM Customs & Excise might consider to be a new motor

Fig 2.5b The valve timing being measured on a one-piece crankcase engine. Note the location of the engine number on the top of the front engine mount. T20 43412 was delivered to C & S Garages, Burnley, Lancashire in June 1958. (The Mike Estall Collection)

cycle, and therefore also avoiding any Purchase Tax liability. The machine, which had been rebuilt as a 1960 model, therefore retained its 1954 registration number. Interestingly, this particular machine still exists today.

2.7 First production machine

The first Meriden despatch book does not show any delivery details for machine T15 101. A very early machine does appear in a June 1953 publicity picture taken in California (see accompanying figure), so it may be that no. 101 was sent there. A Terrier was also shown at a banquet held in the USA for the west coast dealers in February 1953, but whether it was the same machine or another pre-production example is not known.

In common with many other prototypes or experimental machines from Meriden this first production machine may finally have been cut up and scrapped, or perhaps it was sold to an employee and the sale not recorded in the despatch book. Like the other prototypes and the six pre-production machines, its fate is unknown.

The first machine for which any delivery details are known was T15 102. It was despatched to Australia on 24th July 1953, the only machine delivered in the 1953 season.

2.8 'Year' number sequences

Each Triumph production year started and ended at a specific machine number within the continuous number series. For example, the 1957 Tiger Cub season ran from 1st August 1956, with machine no. 26276, to the 31st July 1957, at machine no. 35846. Once the serial numbers reached six digits (at no. 100013), the BSA parent company system came into use at Meriden. Small Heath did not use a single continuous number sequence but each model or group of models had its own number series starting at no. 101 and running up to no. 2000, or starting at no. 2001 or no. 3001 and running upwards.

It has been found that the numbers within each series were not necessarily used in date order and numbers were also duplicated. For example, one group of Mountain Cubs was numbered T20SM 588 to T20SM 709 at Meriden in October/November 1964 and sent to France. A second group of machines, with these same numbers, was built at Small Heath in February/April 1966 and sent to the USA.

Sequences for the machines produced at Meriden went from no. 101 to no. 2000 and those made at Small Heath generally went from no. 2001 upwards. Exceptions were the T20B Bantam Cub numbers and a second series of T20SM numbers, both of which started at no. 101.

2.9 Numbers after 100013 summarized

To sum up and make some sense out of a rather confusing situation, the table on the following page shows what are believed to be the numbering sequences for machines after no. 100013, in approximate date order. These sequences have been deduced from a close examination of the BSA factory despatch records and are thought to be correct. For additional clarity each single sequence has

Fig 2.7 Teenage Californian model Miss 'Bobbie' Borgogno and a Terrier prototype. (The Mike Estall Collection)

been allocated a number by the author. The closing serial numbers can be found in Appendix 2.

2.10 Numbering and model type anomalies. Gaps in number ranges

In all the sequences shown in the table below there are gaps in the despatch books: *i.e.* blocks of numbers apparently not used, except sometimes for perhaps one or two occasional machines. There are also some repetitions where the same sequence of models and numbers has been used more than once and at different times - T20SM has multiple dates and number sequences! Quite frequently the model designation T20SM or T20M may have been stamped on the frame and engine but the opposite model type shown as the model designation in the despatch books. It has been found that there were also

Meriden-built machines	From	Models within each sequence
Sequence 1. no.101 - 2000	September 1964	T20, T20P, T20SM, T20T
Sequence 2. no.101 - 2000	September 1964	T20SC, T20SH, T20SR, T20SS, TR20, TS20
Small Heath-built machines		
Sequence 3. From no.2001	March 1965	T20, T20P
Sequence 4. From no.2001	May 1965	T20SR
Sequence 5. From no.2001	June 1965	T20SM
Sequence 6. From no.2001	July 1965	T20SC
Sequence 7. From no.2001	September 1965	T20 SH
Sequence 8. From no.101	December 1965	T20B Bantam Cubs
	November 1966	T20B Super Cubs
Sequence 9. From no.101	March 1966	T20SM, T20M
Sequence 10. From no.2001	July 1966	T20SM, T20M
Sequence 11. From no.3001	November 1966	T20M

nb1 Machines in sequences 3 to 6 inclusive were assembled at Small Heath but used components made at Meriden.
nb2 The engines in the sequence 5 T20SM machines were built and pre-numbered at Meriden, as were the complete T20SM machines in sequence 1. When T20SM machines came to be built at Small Heath, the same two series of numbers - 101 to 2000 and 2001 onwards - were used, hence the duplication of machine numbers found in the despatch records.

instances where the frame bore one model stamping and the engine bore the other model type!

Then there are some machines shown in the records only as 'Export', the model type not being shown at all. Just to confuse things even further the Mountain Cubs and the French Army Cubs were all stamped in the same sequences of T20SM or T20M numbers. The French Army machines were often (but not always), referred to in the despatch records as T20WD or T20M.WD, but these designations were not stamped on the metal as part of the frame or engine number. The numbers were sometimes suffixed with a letter 'F' after the serial number: *e.g.* T20M 8884F.

2.11 Different frame and engine numbers

One of the Meriden despatch books for 1960 shows two blocks, each of thirty-five machines, where the engines from one block of numbers have apparently been fitted into the frames from the other block of numbers. Therefore, it seems that seventy machines left Meriden with non-matching frame and engine numbers! Originally this was thought to be an example of incorrect stamping of the frame number on the assembly track, but it's now considered more likely that these were clerical errors in the despatch book; the two blocks of engine numbers have been incorrectly recorded against the wrong frame numbers.

This example is by no means unique: many other instances have been found where the frame and engine numbers recorded in the despatch books are not the same, the first being in September 1953 and the last in September 1969. A total of 104 machines are so recorded: *i.e.* 52 apparent 'swaps'.

2.12 Duplication of machine numbers

There are a very few instances where Small Heath has acknowledged that duplication - and even triplication - of serial numbers has taken place. In these cases the records show the serial number followed by a suffix 'A' or 'B'.

Multiple deliveries of machines with the same serial number have been found in the despatch books. These are probably examples of machines which were delivered to a customer on one date, and perhaps returned to the factory. They were then redelivered to another customer at a later date, or even perhaps on the same day. Usually these machines were shown as 'Returned', but as many others are not so shown it's not possible to say whether they were truly returned machines but which were not shown as such, or whether they were actually deliveries of machines with duplicated numbers.

2.13 Differences in model type

The Tiger Cub and Terrier Register (see 2.15 below), which records the serial numbers stamped onto currently existing machines, has brought to light some interesting points. The model type shown in the despatch books did not necessarily correspond with the actual type shown as part of the machine serial number.

The first T20T shown in the Meriden despatch books is no. 56145. The engine to this machine has recently surfaced and has been found to have the number T20S 56145 stamped on the crankcase. The frame and cycle parts have also very recently come to light. The Register has put the owners in touch and this prototype machine is now being restored.

Only six T20W machines appear in the despatch books although it's believed that more were built. The Register has established that at least one of these six machines, shown in the factory records as a T20W, was in fact stamped with an engine number starting 'T20S'. This leads inevitably to the conclusion that perhaps the other five may also have been stamped 'T20S'. Not only that, but perhaps further machines shown as T20S in the records may in fact have been built as T20W models? Apart from the camshaft and carburettor, the T20W was the same technically as the trials variant of the T20S and so the differences between the two models were quite small.

The first few TR20 machines were given engine numbers starting with the prefix 'T20SST', clearly showing the model's derivation as the trials variant of the T20SS. After the first few machines the proper 'TR20' prefix came into being and became part of the normal engine number stamping.

2.14 Financial year & machine numbering. Clerical errors

When it comes to calculating exactly how many machines of each model type were built in any period, there are further complications in addition to differences in model type. There are many clerical errors and omissions within the extant factory records: the only records for the Cub and Terrier now remaining in existence - from which the figures in Appendix 6 have been prepared - are the Meriden despatch books, one Meriden build book and the Small Heath despatch books.

Confusion stems from the fact that some machines were recorded in one book as one particular model and in another book as a different model! Comparison has been made between the one existing build book and its companion despatch book for machines no. 90000 to no. 100013. For the most part the model type recorded against any particular serial number is the same in both books, but occasionally this is not the case. Without a third record it is impossible to say which model type is the correct one! Perhaps in the fulness of time the author's Register may resolve the matter?

These factors have made it impossible now to accurately count how many machines of each type were built. For this reason the figures shown in Appendix 6 must be regarded as best estimates only and not definitive totals.

2.15 Register of machines

An independent record now exists which is gradually throwing light on a very confusing situation. This is the 'Tiger Cub and Terrier Register' which was set up by the author in November 1991. Owners are encouraged to contact the author and tell him the frame and engine numbers of their machines.

The Register's prime function is to record the serial numbers of all currently existing complete machines, also loose frames and engines, to try and determine how many remain in existence. Occasionally, an additional benefit has been the discovery that one owner's frame can be matched with another owner's engine. In these circumstances the author will act as an intermediary, informing each person that the other half of their machine still exists! There are now well over five thousand serial numbers

recorded in the Register, approximately half of which are complete motor cycles (some have been built up from non-matching frames and engines), the remainder being loose frames and engines. Furthermore, the Register is now beginning to reveal some additional detail not apparent in the remaining factory records.

For example, the factory records, particularly those from Small Heath, contain gaps: *i.e.* lines of entry for one machine, or a group of several machines, which have no delivery information. Complete blocks of hundreds, or even thousands, of numbers have also been found unused. The Register is now beginning to show detail in some of those empty spaces, where currently existing machines have been notified to the author and found to belong within them.

The Register has also brought to light another anomaly. For some models, particularly the T20SR, the T20SH and the Mountain Cub, the serial numbers of machines known to exist have reached beyond the highest number shown in the factory records. For example, the highest T20SR number in the records is 2029, but several machines are known to exist with higher numbers. So far the top number in this range is 2080.

At the time of writing 103 machines that were not shown in the factory despatch records are now known to exist. Of those 103 machines, 22 come from the 99,900 numbers listed in the Meriden records and 81 from the roughly 12,800 numbers recorded at Small Heath. The factory records show details of 112,672 machines in all, but added to that figure now must be at least those further 103. The total number of these additions will no doubt further increase as the Register grows, recording the details of yet more currently existing machines not originally shown in the extant factory books.

2.16 New season's colour schemes

Each model and year was allocated individual colours, but these colour schemes frequently ran for several years,

Fig. 2.15a The Tiger Cub and Terrier Register decal. Note the 'T20' pad marking on the right paw - neat, eh? (Designed by Jean Caillou, Paris, 1993)

Fig 2.15b A magnificently restored 1967 Mountain Cub in Grenadier Red and Alaskan White. (Courtesy Dan Andrade, Washington State, USA)

sometimes covering more than one model. For example, both the 1961 and 1962 T20 models were black and Silver Sheen and all T20SH machines from 1962 to 1965 were Hi Fi Scarlet and Silver Sheen. However, it has been found that very late examples of machines from a previous year could be painted with the colours for the next season.

For example, some very late 1960 T20 models, which should have been painted Crystal Grey and black, have been found with Silver Sheen and black paintwork as if they had been 1961 models. Colours for the important American market were frequently different to the home and general export machines, and often anticipated them by a year.

One small touch that demonstrates how keen Edward Turner was on presentation and appearance was the treatment of the wheel rims. All the Terriers and the Cubs, up to the end of the 1956 season, followed the trend set on the contemporary twin cylinder models by having their wheel rim centres painted Amaranth Red with gold lining or Shell Blue Sheen with black lining. It was a small detail which added greatly to the general appearance of the machine. Unfortunately, probably in the interest of cost-cutting, the practice did not extend

beyond the 1956 season.

Something else that demonstrated Turner's concern that the machines should always look good is to be found in the sales brochures. Here, it was the usual practice to blank out all the control cables from the illustrations, thus presenting to the potential buyer a clean-looking image.

The method of colour selection for each model and year is interesting. The design department, in conjunction with the paint shop, would create a whole series of colours and colour combinations and have them sprayed onto petrol tanks for examination by senior management and, in particular, Edward Turner. The tanks would then be laid out for inspection, usually on the lawn at the back of the factory. There may well have then been some discussion on the matter but this usually made no difference in the end: in a perfect example of the democratic process as practiced at Triumph, E.T. would look over the artwork and then inform the assembled company that this model was to be that colour, that model would be this colour, and so on. No argument or discussion was allowed, no contradiction permitted. Autocracy ruled. He had spoken and that was that: end of story.

Fig 2.16 1959 and an experimental black and white colour scheme, or a machine for someone special in the USA? (The Mike Estall Collection)

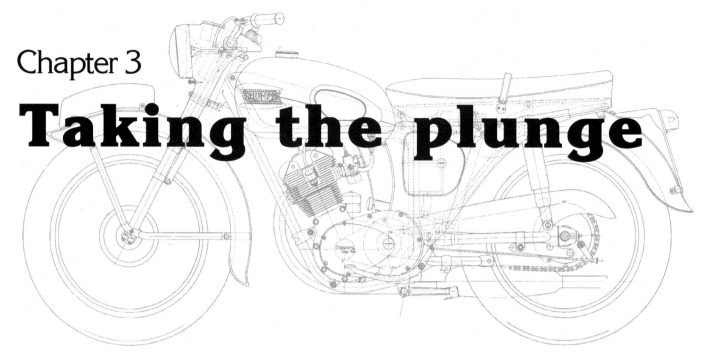

Chapter 3
Taking the plunge

Machines nos. 101 (July 1953) to 26275 (July 1956).

3.1 Plunger suspension

From the start of Terrier production to the end of the 1956 season, the lightweight machines featured plunger rear suspension, Triumph's only venture into that type of springing and entirely appropriate at the time. Some twin cylinder models used the sprung hub but this device was far too heavy and a rigid rear end would have been unthinkable. Swinging fork rear suspension had only seen limited use by other manufacturers at the end of 1952 and Triumph's larger machines did not yet use it, so plungers were the choice. Generally speaking, the plunger system was reasonably satisfactory, although there were a few difficulties which are described later in this chapter.

3.2 Overall gearing

There is no doubt that the Terrier was very fast for its size. The author recalls being told by none other than chief tester Alex Scobie that he had got 'over eighty' out of one of the pre-production machines! This claim may have been the result of a fertile imagination or a well tuned speedometer, but it should be remembered that the very earliest machines had an eighteen tooth gearbox sprocket. This was replaced by a seventeen tooth component very early in the production run, so it is possible that speeds of this order may have been achieved in certain circumstances.

Having a fairly flat torque curve, the engine's overall gearing was set at a level such that, in neutral terrain and weather conditions, the engine did not quite reach peak revolutions at top speed. This was a well intended effort to preserve the engine, but it meant that the machine would inevitably be slightly over-geared. With its original eighteen tooth gearbox sprocket the engine could only 'rev out' in exceptional circumstances, such as

downhill or with a strong following wind. In certain conditions this level of gearing would have been uncomfortably tall. The change to a smaller sprocket in early production brought the peak engine speed and top road speed to a much better compromise.

The new machine was very lightweight, had good brakes and was great fun to ride. Clearly, this was a very desirable package but, in light of later events, the loyalty of new owners was sometimes put to a pretty stiff test by

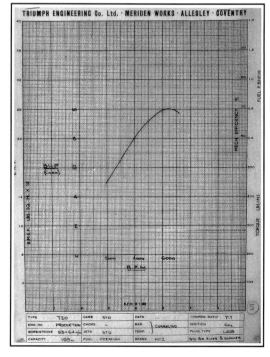

Fig 3.3a The T20 power curve on premium fuel gives a full 10bhp. (The Mike Estall Collection)

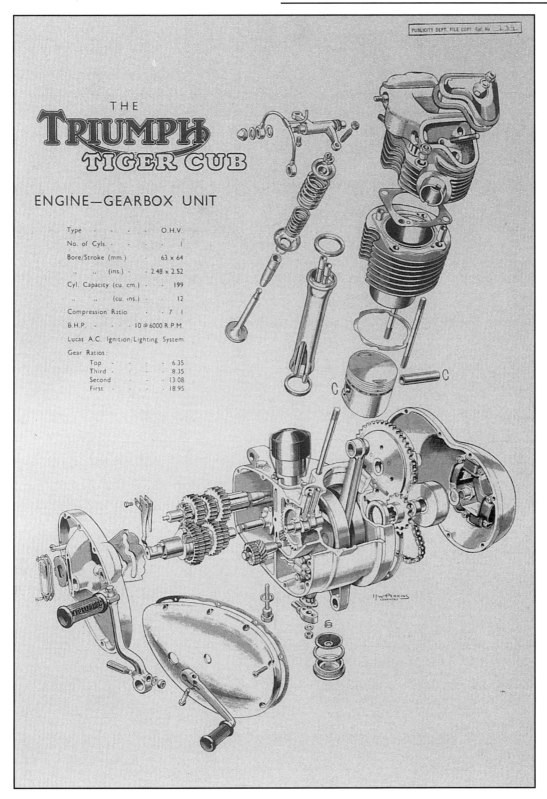

THE

TRIUMPH
TIGER CUB

ENGINE—GEARBOX UNIT

Type	O.H.V.
No. of Cyls.	1
Bore/Stroke (mm.)	63 x 64
" " (ins.)	2.48 x 2.52
Cyl. Capacity (cu. cm.)	199
" " (cu. ins.)	12
Compression Ratio	7 : 1
B.H.P.	10 @ 6000 R.P.M.

Lucas A.C. Ignition/Lighting System

Gear Ratios :

Top	6.35
Third	8.35
Second	13.08
First	18.95

Fig 3.3b The first of four basic engine configurations. One-piece crankcase, distributor, 'Round' head and barrel, simplex primary chain. See also Figs.3.11, 5.12 and 5.14. (Courtesy Trevor Gleadall, JR Technical Publications)

Fig 3.4 Engine compartment detail from a very early 1954 T20. (The Mike Estall Collection)

3.3 The Tiger Cub introduced

In late 1953 the Triumph range was extended with the introduction of a new model, the T20 'Tiger Cub' announced at the Earls Court Show in November of that year. The new model was billed as the sports or 'Tiger' version of the Terrier and, in common with the sports versions of the Speed Twin and Thunderbird (the Tiger 100 and Tiger 110), arrived resplendent in Shell Blue Sheen livery.

The Tiger Cub was much the same as the Terrier in

Fig 3.6a A late 1955 home market T20, viewed from above. Note the separate cable cut-outs in the nacelle - from no. 17258 onwards. (The Mike Estall Collection)

many respects. It had a new flywheel assembly, cylinder barrel and piston with a bore and stroke of 63mm x 64mm, making 199.5cc. The new model had a plain big end from the outset. There was a new cylinder head, a larger Amal 332 carburettor, and one tooth had been added to the gearbox sprocket. A new 80mph black-faced speedometer was fitted and a different gear indicator medallion sat on the nacelle. There were larger section tyres and the mudguards had a central raised rib, whereas those for the Terrier were of plain section.

The Terrier had a saddle as standard but the Cub had a twinseat, each model having the other type as options. The Terrier sported a low level exhaust and the Cub a high level system but with a low level option.

The new T20 engine produced 25% more power than the T15 unit and the Cub was an immediate hit. Top speed went up by about eight or nine miles per hour with very little sacrificed by way of economy, and the acceleration of the Cub, particularly from a standing start, was quite outstanding for its class. In the new colour scheme, with those figures and at a cost of £117.12.0, (£117.60), the Cub was immediately in great demand.

3.4 Tiger Cub - first impressions

The initial response to the new model by the motorcycling press in November 1953 was enthusiastic. The ease of starting was admired and the gear indicator thought very useful. It was also reported to be very quiet and flexible, although a deaf ear must have been turned towards the silencer! The machine was reported to have shown an effortless sixty on the clock and it was said that the factory testers could get another ten on top of that. Acceleration and braking were excellent and handling reported as superb.

3.5 Tiger Cub - delay in early sales

Like the Terrier before it, there was a delay of several months between announcement of the new model and availability for sale; the first deliveries not occurring until March 1954. Whilst a few early machines went to UK customers, the first fifty or so were spread in pairs around the globe, to several European and African countries, Australia and New Zealand, the USA and Canada, Ceylon, South America, and so on. Clearly, the intention was to spread the good news far and wide and as soon as possible.

3.6 Press road tests

An entirely new carburettor by Amal, the 332, had been designed specifically for the Terrier, making a valuable contribution towards the good overall performance. Surprisingly, the Terrier never was road tested in this country but the Australian press reported over 100mpg average fuel consumption and a top speed of around 60mph, depending on conditions.

In the USA a June 1954 Terrier road test by *Cycle* magazine reported a top speed of just over 68mph into a slight headwind, gasoline consumption of 84mpg US, (equivalent to about 105mpg Imperial), and a standing

problems that should have been cured long before.

quarter mile time of less than twenty seconds. The American magazine was almost ecstatic with praise for, 'Its ability to rapidly accelerate from 50mph in top cog .... and anywhere above that figure the rider could feel the stock cams "come in" and push harder....' The author wonders just how standard that particular Terrier was!

In a later Tiger Cub road test the same magazine owned up to fitting a larger main jet and removing the air cleaner, so one wonders if the same tactic had been employed with the Terrier. Or was this dazzling set of figures perhaps due to the unacknowledged first appearance of the 'R' cam, and/or the high compression piston in the supposedly standard machine supplied to the press by Californian distributor Johnson Motors? These reported performance figures were very close to those achieved by the much later 199cc sports Cubs!

The first British road test was of a 1956 T20 checked out by *Motor Cycling* in the 20th October 1955 issue. They gave it a glowing report - 'compact and attractively designed', 'subdued high song of engine and transmission parts working in harmony', 'acceleration from standstill was first class', and so on. The actual performance figures obtained were a top speed of 66.7mph with a fuel consumption of 136mpg at 30mph, 112 at 40 and 90 at 50mph, and a braking distance of 29 feet from 30mph. Acceleration from standstill was given as 3 seconds to 30mph, 6.5 seconds to 40mph and 12 seconds to 50mph. Not bad for 199.5cc.

3.7 Problem areas - general

Despite the nine month wait prior to the sale of the first Terriers, early models were beset with faults, clearly demonstrating that insufficient time and resources had been devoted to development of the new model.

Rectifying the early problems with production Terriers must have been very expensive: not only were owners' machines being returned to the factory to be put right by the Service Department, but production rectification was put in hand, too, reworking areas of the machine before delivery. Two people devoted virtually the whole of their time, for weeks on end, to rectification of design failings that should not have been there in the first place.

The situation had become so serious that, in October 1954, Triumph produced a seven page booklet enti-

Fig 3.6c Front cover of the 'Green 'Un' magazine with a 1956 T20. (Courtesy EMAP)

Fig 3.6b Head-on view of a late 1955 home market T20. (The Mike Estall Collection)

Fig 3.9 An American oil pump with a pressure relief valve. (The Mike Estall Collection)

tled, 'Lightweight Service', listing twenty-three common faults that had arisen in the first year. Some were serious problems requiring major work - for example, a broken crankpin which needed a complete engine strip to repair. Another was a fractured centre stand lug requiring replacement of the entire frame!

The booklet, which also listed oil leaks, heavy oil consumption, transmission problems, fractures of various parts, and so on, told the dealer or distributor how to deal with the problems. Significantly, Triumph found it necessary to apend a note to the front cover of the booklet that said: 'It is not intended that this information should be disclosed to owners'. Turner had boasted that the Terrier was designed in under eight weeks - and this was the result ...

3.8 Problem areas - lubrication and oil leaks

One of the Terrier's main shortcomings was the lubrication system. Overheating problems were being caused by an oil pump and oil tank that were both too small. An inadequate amount of oil was being circulated at too low a rate for this very busy and hot running little engine. It might have been alright on test machines at the factory, but as soon as Joe Public became involved the necessary regular oil changes and level checks were frequently not carried out, so the oil either rapidly ran out or became contaminated and unable to do its job properly.

One problem was that the deep cylinder barrel spigot could get extremely hot, transferring heat to the oil mist inside the crankcase and making what little oil there was work impossibly hard in its capacity as coolant.

The fault was soon acknowledged and the oil tank capacity increased for the start of the 1956 season. The oil pump itself was the subject of continuing changes in size, material and drive mechanısm. The first change came as early as April 1954 and the final version was not made until 1966, by which time it had been found necessary to increase the rate of flow to more than two and a half times the original level! A full listing of all known pumps can be found in Appendix 3.

In September 1954, to stop oil seeping back through the pump and filling the crankcase, and as a back-up for the existing ball valves, a pair of auxilliary spring loaded ball bearings was sited between the pump and the crankcase. These secondary non-return valves lasted for over ten years but from August 1965 were no longer specified.

The early machines were particularly prone to oil leaks. Turner's 'Minimum amount of metal', philosophy made for sealing surfaces that were too narrow, allowing oil to easily migrate across the joint faces to the outside world. The two engine outer covers were also too flimsy, distorting very easily; so, too, were the rocker covers which could buckle under heavy-handed spanner work. All these items were later remade in more substantial materials.

The felt oil seals initially used were ineffective and, as experience was gained, they were gradually replaced by other types. The later use of a rubber labyrinth seal behind the clutch, together with an extended high gear bush, did much to keep the oil inside the primary chaincase. Gearbox seals developed from the felt ring, to one garter seal, then two garter seals fitted back to back, and finally back to one garter seal again.

3.9 Problem areas - excessive oil pressure

By the end of 1958 it became evident to the Americans, with their continuing use of the Cub in its highest performance mode, that big end failures might sometimes be attributable to excessive oil pressure. Normal idling pressures were expected to be around 35 to 50psi, rising to about 125psi at higher revolutions, but a Service Bulletin from the Triumph Corporation reported that the pressure could sometimes exceed 200psi!

The Americans reported to Meriden their opinion that these very high pressures were somehow contributing to the high rate of big end failure, but this was not taken seriously and nothing was done. However, in the USA at least two after-market manufacturers (Webco and Bates) produced oil pumps or parts thereof which provided adjustable pressure relief. The Bates item was a replacement oil pump feed plug with an adjustable spring loaded valve which cost $1.95 in 1959. Webco made a complete replacement pump which in 1961 cost $18.50. Both solutions maintained oil pressure at any constant preset level at all higher engine speeds.

3.10 Problem areas - 'wet-sumping'

Lack of oil pump scavenge could sometimes produce excessive crankcase pressure, wet-sumping and severe incontinence. The condition was typified by the crankcase rapidly filling with oil, which would then blow out of the breather tube at the back of the crankcase, also forcing its way out of every other joint in the engine cases. At the same time the engine would labour, working hard to overcome the excessively high pressure under the piston.

As well as dirt or wear in the scavenge side of the oil pump, or an obstructed breather, it was found there was another major reason for the problem. With the advent of the two-piece crankcase came the possibility that the oil return pipe could lose its prime: *i.e.* it would suck air rather than oil. This was most likely to happen if an engine was subjected to continuous use at high speeds in combination with high ambient temperatures, conditions frequently encountered in competition in the western USA.

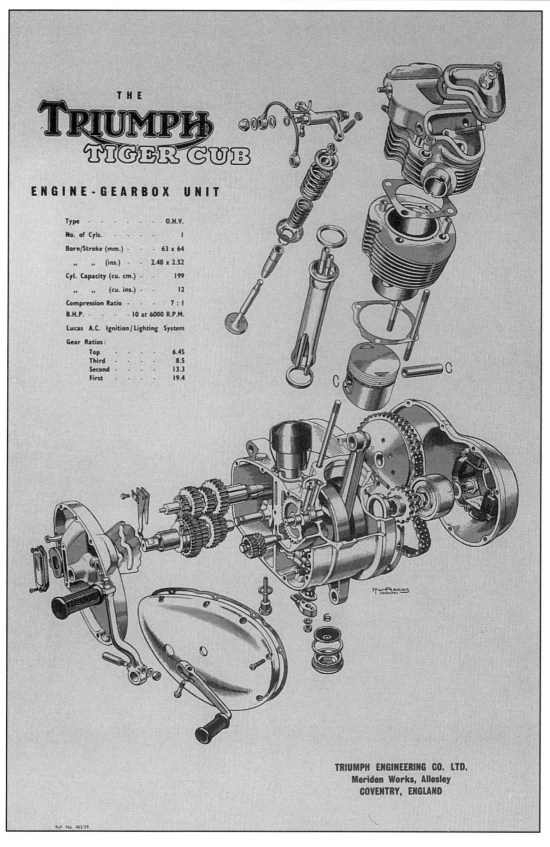

THE

TRIUMPH
TIGER CUB

ENGINE - GEARBOX UNIT

Type	O.H.V.
No. of Cyls.	1
Bore/Stroke (mm.)	63 x 64
„ „ (ins.)	2.48 x 2.52
Cyl. Capacity (cu. cm.)	199
„ „ (cu. ins.)	12
Compression Ratio	7 : 1
B.H.P.	10 at 6000 R.P.M.
Lucas A.C. Ignition/Lighting System	

Gear Ratios:
Top	6.45
Third	8.5
Second	13.3
First	19.4

TRIUMPH ENGINEERING CO. LTD.
Meriden Works, Allesley
COVENTRY, ENGLAND

Ref. No. 463/59

Fig 3.11 The second engine configuration. One-piece crankcase, distributor, 'Oval' head and barrel, duplex primary chain. See also Figs 3.3b, 5.12 and 5.14. (Courtesy Trevor Gleadall, JR Technical Publications)

The problem was also a function of the design of the oil pick-up pipe. The inside diameter of the pipe was found to be too large to properly collect hot, thin oil from the scavenge well in the crankcase bottom. Lack of scavenge could even occur when the pipe end was completely submerged in lubricant while the machine was running on a smooth road. If the machine was bouncing around over bumpy terrain while in competition then it was a near certainty. Having recognised the problem the factory supplied a kit to cure it. This consisted of a new supplementary scavenge pipe in the form of a sleeve that fitted inside the existing item, thus reducing its diameter, and also lighter oil pump auxilliary ball springs. However, this sleeve could not be used with pick-up pipes that had an angled end-cut. These pipes had to be removed from the crankcase and a new smaller diameter, straight-cut pipe substituted.

A change in pump material from brass to cast iron in 1962 was also said to give better pump scavenging at higher temperatures. This may well have been the case but the oil pump material reverted to brass again in its final incarnation with the Mountain, Bantam and Super Cubs from 1966. The main reason for the change back to the brass material was that the ball seats in the cast iron pump tended to become oval after extended use.

3.11 Problem areas - clutch and primary drive

Clutch modifications were legion: the number of plates, number of teeth, chain size, clutch wheel design and construction, friction material and bearing type all changed. Originally, both the Terrier and Cub had two drive and two driven plates, but the drive plate tags tended to curl up or roll over when the full eight or ten horsepower was unleashed, so by February 1955 the number of plates had been increased to three of each type.

At the same time the clutch centre driving cups changed from pressings to being machined from the solid. The backing plate drive pin material was improved and the pins were given longer threads, which were peened over at the back of the plate.

It was soon found that the clutch cable was too lightly constructed and was stretching. There was also no provision for adjustment in the pressure plate and insufficient leverage in the clutch control lever, so the clutch action gradually became less effective as cable slack increased. By the middle of 1954 all of these faults were corrected.

In the beginning both models had a toothed ring clutch sprocket with a single row of 48 x ^{3}/$_8$in x ^{7}/$_{32}$in teeth. There were two versions of this sprocket as, after excessive wear was found, the sprocket had to be hardened. The toothed ring sprocket then changed to a pressed basket design with a riveted-on single row sprocket of 48 x ^{3}/$_8$in x ^{7}/$_{32}$in teeth. For the Cub this item changed to 36 x ^{1}/$_2$in x ^{3}/$_{16}$in teeth in August 1955.

The Terrier's pressed basket design then became a casting with an integral single row of 48 x ^{3}/$_8$in x ^{7}/$_{32}$in teeth, and this was the last design for the simplex chain. The final design change was the duplex type, for the Cub only, with a double row of 48 x ^{3}/$_8$in x ^{7}/$_{32}$in teeth.

The primary chain and engine sprocket sizes also changed in line with the clutch wheel alterations, going from 19 teeth x ^{3}/$_8$in x ^{7}/$_{32}$in simplex to 18 teeth x ^{1}/$_2$in x ^{3}/$_{16}$in simplex then finally back to 19 teeth x ^{3}/$_8$in x ^{7}/$_{32}$in, this time of duplex configuration. The ^{1}/$_2$in x ^{3}/$_{16}$in size was a special endless chain made to very close tolerances in the hope that it would maintain tension. Even then there were several premature failures and it was not until the advent of the duplex type that a satisfactory solution finally arrived.

The clutch friction material had changed in February 1955 from cork inserts to bonded Neo-Langite segments. This material remained in use right up to the 1965 season, when it was retired in favour of a granulated cork compound with some rubber content.

The clutch bearing started off as a riveted-up ring of 58 x ^{5}/$_{32}$in balls, but this changed to a plain cast iron slipper ring in April 1954, changing again in July that year to a plain phosphor bronze slipper ring. Almost immediately the bearing changed again with the addition of 20 x ^{5}/$_{32}$in balls to the phosphor bronze ring, and in February 1955 the final type, with 16 x ^{5}/$_{32}$in balls, arrived.

Most of these changes took place in the period up to the end of the 1957 season and many early machines even had to be removed from export crates to have the transmission reworked. But by the start of the 1958 year the clutch and primary chain difficulties had been overcome and the primary transmission had settled into its final form.

3.12 Problem areas - electrics

One of the electrical problems was caused by insufficient wire in the wiring loom where, on full steering lock, wires could be pulled out of the terminals in the back of the Lucas PRS8 switch. From machine no. 8120 a cable clip was fitted at the front petrol tank mounting bolt to prevent pulling on the wires. At the same time the inclusion of snap connectors in the nacelle gave the wiring a little more effective length, also making it much easier to remove the nacelle top, a task which could now be accomplished without first detaching the switch.

The distributor drive shaft had no direct means of lubrication until July 1960. Any wear in the distributor bush would cause sideways movement in the shaft, which would be transmitted up to the contact breaker, altering the gap setting, the ignition timing, and possibly causing other problems.

From engine no. 69157 a new timing side main bearing bush, which had a slot in its periphery, came into use. The slot allowed oil to bleed to the skew gears, also providing residual lubrication upward into the distributor bush. Owners were made aware that the slotted bush was not to be used on machines with the earlier, small oil pump as diversion of oil to the skew gears could leave an inadequate feed for the big end.

The complicated wiring harness used with the PRS8 switch was difficult to refit correctly once taken off the machine. With eighteen terminals in the back of the switch, some of which had two or even three wires crammed into them, the result was the electrical equivalent of a plumber's nightmare. The possibility of erroneous re-connection, producing non-functioning electrics or damage to the components, was high.

.The associated wiring diagram was also unclear around the switch area and, over the years, several changes to the diagram only confused matters further. The PRS8 switch also had a very fine thread on its top half which could easily be stripped or crossed by a ham-fisted owner. Worse still, the top of the switch could be completely ripped off by excessive force with a spanner.

When refitting the rectifier, overtightening of the centre bolt could cause damage, giving rise to charging problems. Also, early germanium components proved untrustworthy and the change to a silicon material greatly improved reliability.

Alternator rotors could work loose, damaging the keyway. The resultant clattering of the rotor on the engine mainshaft sounded just like serious big end wear. Later there were problems with the Energy Transfer system used on some sports machines, but this system did not come into use until 1958 and more detail is given later.

3.13 Problem areas - rear suspension, centre stand, frame, cables, rear chain

The cycle parts were, on the whole, less troublesome. The torque stay peg on the left side rear plunger was initially only a press fit and sometimes pulled out of its housing. It was later replaced by a screw-in peg. Another problem was breakage of the light alloy fork ends, but this was infrequent.

The rear suspension units required - but seldom received - regular greasing. Without the essential lubricant they could go on strike completely, seizing solid and effectively turning the machine into a 'rigid' mount. Excessive wear between the hardened steel plunger guide rods and the alloy fork ends in which they ran could be a problem, too. It not only made the rear end of the machine wobble like a jelly as the wheel flopped about, but if in combination with a worn or slack chain, divorce between the sprocket and chain was the likely result. The chain would then jam between brakeplate and suspension cover and, if the rider was really unlucky, the rear wheel could lock up, too - with dire consequences.

The early tubular centre stand did not survive long, particularly among owners who kick-started the bike while still on the stand: it was soon replaced by a much stronger forged item. The frame lugs that carried the centre stand also proved too weak; when a lug broke the only solution was to replace the entire frame! An improved lug was brought into use at machine no. 2278 and the problem was resolved.

Initially there had been no side stand, so the poor centre stand had to do all the work, but a prop stand lug was fitted to the frame from early 1954 with the stand as an optional extra. This particular weakness, and also control cables that were of too light a gauge, were commented upon in the 1954 *Cycle* magazine road test in the USA.

One major problem had centred on the front frame loop where rough riding could induce the 'swan neck' to flex, sometimes to the extent of splitting the petrol tank. The fuel tank had been intended to give some light support in this area, but was not tough enough. Meriden soon acknowledged the problem and, from August 1955, this area was triangulated by putting pressed steel plates right through the tank. These plates replaced the welded-on tank mountings and braced the headstock.

3.14 Problem areas - big ends and flywheel assemblies

The big end was another story. The first Terriers were fitted with a 1 1/2in diameter roller bearing made in-house at Meriden. The fifteen rollers, although uncaged, were nonetheless retained within the bearing by tracks machined in the crankpin. The problem here was that Triumph could not make the track and rollers to the required degree of accuracy or parallelism, with the result that the rollers tried to crowd over to one side of the track, rubbing their end faces against the containing walls of the crankpin track and putting the whole bearing in jeopardy. The cure would have involved selective assembly of components but this was not feasible under current production conditions.

By April 1954 the Terrier roller bearing had been replaced by a 1 1/8in diameter steel-backed white metal bush. The crankpin spigot size into the flywheels remained for the moment at 3/4in. The plain white metal big end was used on Cubs right from the outset. The bush size was adequate for normal use and it lasted right through to 1962, but the bush material was changed from white metal to the steel-backed copper-lead Vandervell VP3 type in August 1956.

A further type of big end assembly had fourteen needle rollers which were retained by a steel cage. Here, the crankpin track diameter was slightly larger than the 3/4in spigot into the flywheels. This type is not shown in any factory literature and it's possible it was made by an external sub-contractor.

To add yet more strength and rigidity to the flywheel assembly, the spigot size of the crankpin into the flywheel was increased from 3/4in to 13/16in in August 1955, but still using the 1 1/8in diameter plain bush. Another big end design was produced by Alpha Bearings as an after-market replacement. This type of big end had ten rollers contained in an alloy cage running on a steel track in the conrod eye, and had a 13/16in diameter crankpin.

Then came an increase in the diameter of the big end bush from 1 1/8in to 1 5/16in during February 1962, and also a two-piece crankpin. In August 1964 a 'Deva Metal' bush was tried and the one-piece crankpin returned. This bush was made of a sintered phosphor bronze material impregnated with graphite, but it was not very successful. The big end continued to give trouble until, finally, a steel-caged, single row, needle-roller big end bearing came into use in August 1966. This type of bearing was at last completely up to the job and no further problems arose, but by then it was almost too late for production was to cease within a couple of years.

Over-enthusiastic tightening of the bolt holding

the skew gear and timing pinion onto the end of the flywheel assembly caused problems. A small peg ensured the pinion was accurately positioned and, as the bolt was tightened, the pinion was drawn onto a taper machined inside the flywheel stub. Unnecessary and excessive force on the bolt caused the taper to spread, sometimes cracking the flywheel stub around the peg hole.

Once this had happened, any hope of maintaining good oil pressure to the big end was gone; not only that, the bolt itself could be stretched by heavy-handed spanner work. If that happened the bolt end could protrude into an oilway, interrupting oil feed to the big end. It was a neat design, but in the real world of home maintenance insensitive use of a spanner could do untold damage!

The combination of the Terrier and Cub stroke dimensions, various big end types and sizes, two timing side main bearing types, different drive side mainshafts and two thicknesses of flywheel meant that, over the years, nearly twenty different types of flywheel assembly had been produced.

3.15 Problem areas - troubles due to performance

As soon as the public got hold of the new machines the problems really began. The Terrier was no slouch for a 150cc machine and the Cub in particular suffered as a direct consequence of one of its major assets - outstanding speed and acceleration. Quite simply, it had too much performance for its own good.

The Cub could burn off any scooter, most lightweights and many machines of much larger capacity. Young and inexperienced riders (such as the author), would wring the bike's neck with great enthusiasm, squeezing out every last ounce of performance until something gave way! Not being a very robust animal, this happened with distressing frequency.

The Cub was particularly good leaping off the mark at traffic light Grands Prix, putting much larger machines to shame. With mechanical clatterings from the engine room and a prominent exhaust note, the Cub always somehow seemed to its rider to be going fast and young lads everywhere could be seen lying flat on the tank with one eye on the horizon and the other on the speedo! On an engine that had not perhaps been thoroughly warmed up and which maybe had not seen an oil change since leaving Meriden, it was a recipe for disaster.

The engine unit was not robust enough to put up with continued treatment like this and the failures came thick and fast: big ends went bang, main bearings grumbled, pistons screamed in seizure and the early clutches threw themselves to pieces. The author's 1956 Cub, while being subjected to a particularly prolonged and inconsiderate handful of throttle, seized its piston and actually split the crankcase right along the top, from the back of the cylinder barrel plinth to the back of the gearbox. It cost a week's wages to repair!

Thus was revealed another weakness. The crankcase itself was not rigid enough and could flex under extreme conditions, so adding to the list of problems.

Fig 3.15 An early 1956 export machine. Note the bolt-on handlebar levers and oil tank filler cap position - still in the 1954/5 location. This photograph might be of a styling mock-up machine. (The Mike Estall Collection)

Once again Edward Turner's philosophy of making the minimum amount of metal do the maximum amount of work was too idealistic. Still, it did have one indirect benefit. Frequently, the owner of a by now moribund Cub, would decide that, despite everything, he still liked Triumph motor cycles and wanted another one. His next move would be to ditch the lightweight in favour of a big twin! Thus was born marque loyalty - perhaps E.T. wasn't so wrong after all?

3.16 Problem areas - noise

To the ears of the average young enthusiast the Cub was endearingly loud! Mechanical noise on such a small, hard-working engine was always going to be difficult to suppress and the general decibel level could be satisfyingly obtrusive. The Cub and Terrier also had a very characteristic exhaust note which could be heard and recognised above and beyond almost any other machine. On the early models in particular it was an unpleasant, hard, flat crack, audible streets away from the source.

Being such fun to ride, the average young blood could not resist winding up the twistgrip and, in any case, riding a Cub quietly was nigh on impossible. The author very soon ditched the silencer on his own 1956 Cub and substituted one from a G9 Matchless, suitably modified to fit the pipe. Not only was it megaphone-shaped, generating illusions of racing grandeur, it also gave out a much deeper booming noise. It made the bike sound much bigger and faster and was also very much easier on the ear!

3.17 Problem areas - other

One other minor fault concerned the 'spectacle' disk originally fitted inside the pushrod tube to assist in location of the pushrods. In early engines the pushrod tube was positioned by a small peg at the top of the crankcase. If that peg came out, or if the owner did not locate the tube properly, the disk would carve chunks out of the pushrods. In later engines the peg was omitted and the pushrod tubes were made without the disk. Owners were advised to punch the disk out from the pushrod tube whenever the opportunity presented itself.

3.18 Problem areas - summary

It sounds as if the whole thing was a catalogue of disasters: not so! Certainly, many parts of the machine were under-engineered from the outset and suffered ills that could have been prevented if component construction had been more substantial. However, with normal use by owners who took some care and interest in the well-being of their mounts, the Triumph lightweights would, in most cases, perform admirably and reliably. They were fast, very economical and handled well.

Problems arose mainly because young and inexperienced riders pushed the little engines too hard, trying to use them beyond original performance intentions and design parameters, until something broke. The lightness of construction had stemmed directly from Turner's 'minimum amount of metal' philosophy, but gradually

Fig 3.19a A 1956 T20 with a Terrier just visible behind it. Both machines were restored by the author. (The Mike Estall Collection)

Fig. 3.19b Another shot of a 1956 T20. (The Mike Estall Collection)

the problems were ironed out and reliability slowly improved.

Tens of thousands learned to ride and served their motor cycling apprenticeships on these endearing little machines, and many grew to love them. Over the years the machine gradually matured and improved until, finally, by the time it was right, Triumph stopped making it!

3.19 Total sales of the plunger models

There was a nine month delay between announcement of the Terrier in early November 1952 and the first recorded sale of a production machine. This was T15 102 which was sent to Adelaide, Australia on 24th July 1953. Further machines dribbled out of Meriden during the months of August and September with deliveries picking up from October 1953. The Tiger Cub was first exhibited at the November 1953 Earls Court Show, just one year after first sight of the Terrier, but UK deliveries did not start until March 1954.

The factory despatch books show a total of 9237 Terriers and 16,212 plunger Cubs being delivered in the 1954 to 1956 seasons. The Terrier figure includes 30 machines delivered to Indonesia in 1957. The total of 25,449 machines compares with just under 39,000 twin cylinder models delivered in the same period.

3.20 Fochj - 'The Italian job'

One interesting development that never went past the experimental stage took place in Italy during 1955 with a company named 'Fochj', from Bologna. In the mid-fifties this firm had marketed lightweight motor cycles assembled from bought-in parts, usually German NSU engines in Italian cycle parts. They were purely assemblers of machines, manufacturing little or nothing 'in-house'.

They must have liked the idea of a model with the Cub or Terrier engine for they built a small number of prototype machines for evaluation. First was a Terrier-engined roadster and then a Cub-engined sports machine. The engines were standard Triumph units, except for some additional finning on the sparking plug side of the cylinder head, and had a continental 'heel and toe' type of gear lever. The cycle parts were Moto Guzzi mudguards and fuel tank, perhaps a Mondial or Guzzi swinging arm frame and Ceriani front forks with Ducati front and rear hubs. Two of these interesting machines were exhibited at the November 1955 Milan Show but the project, for whatever reason, was stillborn and, as far as is known, no production models were built.

3.19c & 3.19d A 1956 T20 in Meriden village. Just one mile from the factory and Meriden is said to be the 'Centre of England'. These photos of T20 17513 were taken on the 1st September 1955. (The Mike Estall Collection)

Fig. 3.20a An example of a Terrier-engined Fochj motor cycle – from the timing side. (The Mike Estall Collection)

Fig. 3.20b The Fochj project - from the drive side. Note that it says 'Terrier' on the primary cover but '200cc' on the tank. (The Mike Estall Collection)

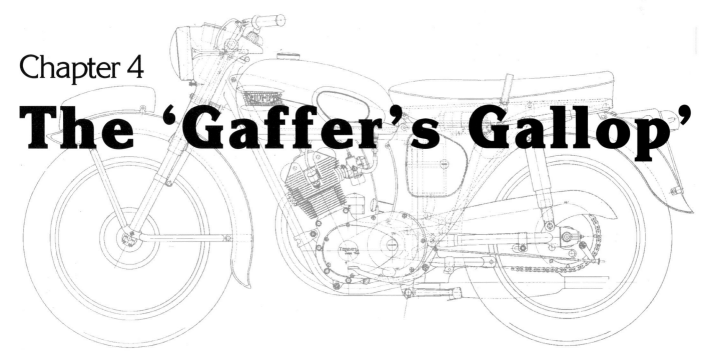

Chapter 4
The 'Gaffer's Gallop'

4.1 The runners and riders

Early difficulties with the Terrier had given rise to concern amongst many Triumph dealers and, if not addressed, sales could have been seriously affected. To allay dealers' fears - and to demonstrate the reliability and economy of the 150cc lightweight - the factory dreamed up a publicity stunt: three senior staff would ride three Terriers from Lands End to John O'Groats, visiting on the way as many Triumph dealers as possible, and with the overall target of covering 1000 miles at an average speed of 30mph with 100mpg fuel consumption. This exercise was very soon dubbed 'The Gaffer's Gallop'.

The three riders were none other than Edward Turner (Managing Director and Chief Designer, aged 52, on T15 no. 475), Bob Fearon (Works Director, aged 46, on T15 no. 476) and Alec St. John Masters (Service Manager, aged 56, on T15 no. 477). The press was given a handout which described the three as 'The man who designed it, the man who made it and the man who will service it'.

4.2 The route

The ride started at Meriden on Monday 5th October 1953 with an 11am press reception followed by an inspection of the Terrier assembly track, and then refreshments. The three riders and machines left for Exeter at 1.00pm, arriving 7.05pm to spend the night at the Imperial Hotel.

On Tuesday 6th October the riders left Exeter at 9.00am and went, via Okehampton and Launceston, to the Victoria Hotel, Roche, where the test officially started at 11.40am. From there the route was via Redruth and Penzance to Lands End for lunch, returning over the same route to Exeter and the comfort of the Imperial Hotel, arriving at 6.35pm, a distance of 180 miles.

The following day the three left Exeter and pro-

Fig. 4.1 The route map from the official handout. (The Mike Estall Collection)

ceeded via Honiton, Shepton Mallet and Bath, with lunch at the Old Bell Hotel, Malmsbury. Then on to Cirencester, Stowe in the Wold and Warwick, finishing at the Regent Hotel, Leamington Spa at around 4.05pm. This route added a further 163 miles.

Fig. 4.2d Fuelling under the watchful eye of John McNulty from the ACU. (The Mike Estall Collection)

Fig 4.2a It's big smiles from everyone at the start: will they still be smiling at the end?

Fig. 4.2b The list of runners in the 'Gaffer's Gallop.' From left to right - Frank Griffiths (chauffeur), Bob Fearon, Colin Swaisland (Esso cameraman), Steve Tilley, John Dipple (Lucas), Ernie Nott (Experimental Dept.), Edward Turner, John McNulty (ACU), Eric Headlam, Alec Masters. (The Mike Estall Collection)

On Thursday the riders turned north for the first of two very long days. The first covered 246 miles, leaving Leamington Spa at 8.00am, through Coventry, Leicester, Six Hills, Bawtry, Doncaster and lunching at Boroughbridge. From there to Scotch Corner, tea at Applebey, Penrith and finally the Crown and Mitre Hotel in Carlisle by 6.00pm.

Friday saw the longest leg of all - 263 miles. Leaving Carlisle at 8.00am, the trio went via Lockerbie, Lanark and Stirling, with a one hour lunch break at the Salutation Hotel, Perth. Then through Pitlochry and a half hour stop for tea at Dalwhinnie just after 4.00pm, through Kingussie and on to the Caledonian Hotel in Inverness by 6.30pm. Nearly there!

For the final leg on Saturday 10th October they had to leave Inverness at 7.00am (that must have been an effort) and ride on through Dingwall, Bonar Bridge, Helmsdale and Wick to arrive at John O'Groats just

before midday, returning to Wick at 12.50pm and totalling 158 miles.

The riders returned by a British European Airways flight to Elmdon, Coventry, arriving at 8.25pm on Saturday. The machines came back to Meriden by van, stopping at Pitlochry on Saturday night, Doncaster on Sunday night and arriving at the Triumph factory on the morning of Monday 12th October. Here, the machines were stripped and inspected under the watchful eye of the ACU official observer.

Throughout the run the riders were followed by a posse consisting of a car driven by Frank Griffiths, with Eric Headlam in charge of the test, Reeves Quann the Triumph press agent and Jock McNulty, the ACU man. On top of that there was a service van with a mechanic, several hangers-on and a Lucas technician riding a model 5T Speed Twin.

4.3 The performance

The three machines were standard Terriers taken straight off the production line, but no doubt subjected to a little more than the standard post-production inspection. They had to carry Turner's 14 stone (89kg), Bob Fearon's 16½ stone (105kg), and Alec Masters' 12 stone (76kg).

The plan had been to average 30mph and 100mpg for just over 1000 miles. In fact, they achieved an average speed of 36.68mph and an average fuel consumption of 108.6 miles per gallon. Oil usage was an average of 6oz (170gms) per machine over a distance of 1008 miles. Pretty good for 150cc machines carrying a full load, and in the days before the British motorway had been invented!

4.4 The results

The demonstration showed that, given proper servicing and decent treatment by its rider, the little Terrier was a machine capable of some remarkable performance figures. Turner had wanted to produce a machine that was ideal for the novice rider and he had certainly done so. Order books for this model, and the twins, as well, were full for the next twelve months. News of the achievement went

Fig. 4.2c Refuelling being carefully watched by the ACU man. Check the fuel price - 4/5½ (£0.22). (The Mike Estall Collection)

Fig. 4.2e Crossing the border into Scotland. (The Mike Estall Collection)

Fig. 4.2f Only another 300 miles to go! (The Mike Estall Collection)

Fig. 4.3 Alec Masters has obviously just made some wisecrack! (The Mike Estall Collection)

as far as the USA where the Triumph Corporation, on 29th October, put out a newsletter to eastern Triumph dealers notifying all and sundry of the details and confirming that the Terrier was definitely a machine to be reckoned with.

Given the initial set-backs and delays in production, this exhibition of the Terrier's capabilities can have done nothing but good. The riders also proved their mettle: of the three only Alec Masters was a regular motor cyclist; indeed, Bob Fearon could barely ride at all. He and Edward Turner were by no means hardened to the saddle as both were desk-bound during their working days.

The whole test went extremely well. The high average speed achieved in those pre-motorway days was no doubt due to the cracking pace set by E.T. employing his racing skills. No doubt Alec Masters enjoyed himself too, but poor old Bob Fearon had to grit his teeth and hang on. At one point he failed to negotiate a Scottish bend and left the road at some speed, bouncing a hidden culvert before regaining control of his machine and his composure, fortunately without significant damage.

The weather during the run had been all but perfect with some rain and blustery winds in the very north of Scotland only. It did them all credit that they came through unscathed, although no doubt they were a little saddle-sore at the end!

In addition to black and white photographs of the run (photographer unknown), Colin Swaisland took a cine camera and filmed the run in colour for Esso Petroleum. Regrettably, the whereabouts of any copies of this film are unknown, as are copies of the bound book of black and white photographs produced for each of the riders.

One interesting footnote was that Edward Turner used his considerable personal influence to good effect and apparently did not ride his machine the whole way. He occasionally retired to the comfort of his own accompanying Sunbeam Talbot 90 car, allowing its chauffeur, Frank Griffiths, or the mechanic, Steve Tilley, the privilege of doing a spell on the Terrier! It is not known whether the other two riders were accorded the same privilege!

Chapter 5
Full swing at Meriden

Machines nos. 26276 (August 1956) to 100013 (September 1964) and 101 (September 1964) to 2000 (October 1964)

5.1 Design changes - swinging fork frame, steering lock

August 1956 saw the start of the 1957 season and with it came an entirely new frame. As early as December 1954, after-market swinging fork rear suspension conversions had been available, notably from Bill Martin in California, who could also provide Earles-type front forks. Henry Vale at Meriden had built a swinging fork Cub for George Fisher to ride in the 1956 SSDT, and Ken Heanes' ISDT machine also used this type of rear suspension. Although these were 'works' prototype machines, the writing was on the wall for the obsolescent plunger frame.

Experiments had also been made at Meriden on a twin cylinder machine fitted with an Earles-type leading link front fork from an Ariel machine, but there was a non-productive weight penalty. As a telescopic fork rises and falls in its travel, the wheelbase of the motor cycle alters by as much as three inches. This causes the peripheral velocity of the wheel to change rapidly as well, transmitting these changes as loading shocks to the hands and shoulders of the rider. With the Earles-type of fork the wheelbase changes were very much smaller - as were the resulting loads on the rider. However, this fork was much heavier than the telescopic type, a penalty outweighing any advantage gained .

The old plunger frame had been a one-piece item complete in itself, but the new swinging fork chassis consisted of several different frame parts bolted together. The front loop held the steering head, engine mountings and seat tube. Bolted to the front frame was a separate rear sub-frame which formed the seat-carrying loop and carried the rear suspension top mounts. The swinging fork

held the lower rear suspension and chainguard mounting points. There were also two large bolted-on brackets, further bracing the swinging fork area and holding the pillion footrests and prop stand lug.

In mid-1957 came the introduction of a steering lock, at first on the T20C and then the T20. In use, withdrawal of the key left the Neiman lock in place with its extended tongue engaged into a slot milled in the

Fig. 5.1 The Neiman steering lock and housing. (Courtesy John Nelson)

steering stem. The lock housing became part of all frames but the lock itself was an optional extra.

5.2 Design changes - hydraulic front fork

In August 1956, with the advent of the new frame came also a new front fork. Until then the front forks had been grease-filled, devoid of any damping, and in cold weather they became very stiff in action. With grease-filled rear plungers it meant that the rider could be relying only on the balloon effect of the tyres for any degree of comfort. If regular greasing was neglected then any vestige of suspension movement could almost cease altogether.

This rather primitive arrangement came to an end at the start of the 1957 season with the introduction of the new hydraulically damped front fork. This had a much better action and, when taken together with the new swinging fork rear suspension, gave a much greater degree of rider comfort.

The new front fork stanchions were 0.010in smaller in diameter than the old type and were not interchangeable - unless the middle lug and stem was changed too! The top fork bushes were also changed to fit the new stanchion diameter. This dimension change arose because the rubber seals for the hydraulic fork required a more smoothly ground stanchion finish than had been necessary with the felt seals used on the old grease-filled fork.

5.3 Design changes - two-piece crankcase. Experimental paper oil filter

Our flexible friend, the one-piece crankcase, disappeared in October 1959 to be replaced, from engine no. 57617, by a two-piece design. Here, the joint between the two crankcase halves became conventionally placed on the cylinder barrel centre line. A reason for the change, quoted at the time, was: 'to streamline manufacturing methods and to reduce production costs'. This excuse was somewhat lame for it acknowledged that the old one-piece design had been less than satisfactory! The new engine casings gave a much needed boost to overall rigidity of the engine unit.

The new crankcase halves came in matched pairs, and were jig-bored through the main bearing housings. A mandril was then put through the two halves for the cylinder barrel base to be machined, ensuring that the base was flat, the same height on both sides and in the same plane as the crankshaft. The two crankcase halves were then number stamped underneath to ensure an original pair stayed together.

There were also new designs for the primary and timing side outer covers and the contact breaker now gained an internal clamp which replaced the earlier rather untidy external item.

The first machines to be fitted with the two-piece crankcase were a batch of five hundred T20S models intended for the USA. Although the new cases had the virtue of being much more rigid than the one-piece type, there was a disadvantage. If the gearbox sprocket needed changing the entire engine unit had to be completely stripped. Sprocket replacement had previously been relatively simple, only requiring removal of the clutch, primary chain and alternator stator, allowing the drive side inner cover to be removed, thus exposing the sprocket.

With the advent of the two-piece crankcase, access to the gearbox sprocket could only be gained by splitting the crankcase halves, which meant that the cylinder head and barrel had to removed first. The gearbox high gear, bearing, oil seal and sprocket became the very last items to be removed from the crankcase when stripping the engine, and the very first to be fitted when rebuilding.

If a change of gearing was required in competition

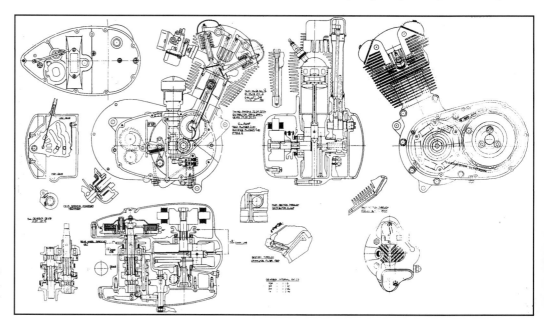

Fig. 5.3a Drawing of a 1960 T20 engine with the two-piece crankcase. (Courtesy John Nelson)

Fig. 5.3b The experimental oil filter - outer cover. The engine is no. 63233. (Courtesy Alan Halford)

then changing the gearbox sprocket became the very last option to be considered, unless a complete engine strip was contemplated too! Much later the Cub-derived BSA C15 model gained a removable cover plate behind the clutch that made the job quite simple. If only the Cub itself had had the same facility!

In early 1960 three experimental engines were built to evaluate effectiveness of placing an oil filter in the feed line to the crankshaft. A tubular paper element filter, approximately two and a half inches long and about half that figure in diameter, was located directly above the oil pump inside a perforated metal cover. It was contained within a small housing bolted through its centre to the crankcase and which protruded through the inner timing side cover. The outer cover featured an additional small cover secured by three screws which could be removed to gain access to the filter. The oil pressure test hole on these three engines was blocked, being neither drilled nor tapped.

Oil was taken from a modified internal feed line and sent through the filter by the normal oil pump action before being returned to the crankcase oilway and onwards to the big end. The outcome of the experiment is unknown but this feature was not taken into production, perhaps being considered too costly. Had it done so it might have gone some way towards creating a longer life for the engine unit, dispelling the Cub's well-earned reputation for self-destruction.

5.4 Design changes - general improvements

The timing side main bearing bush was replaced by a ball journal bearing in February 1962. The contact breaker 'mushroom' disappeared in August 1962, giving way to the 'side-points' engine. Engine cooling was gradually improved by increased finning on the cylinder head and barrel and by the use of larger oil pumps. Big ends were made larger and generally more robust. Carburation was the subject of a whole series of changes. The clutch and primary drive problems have already been described and by late 1957 a cure had been effected.

The cottered kickstart lever was replaced by a splined type for 1965 on most of the sports models. It had a

kickstart carrier which swung the lever outwards and clear of the exhaust system. It was not a great success as the carrier was not strong enough to deal with the increased leverage exerted by the new pedal design, and the original cottered lever reappeared for the 1966 season.

5.5 Design changes - silencing

Silencing was an issue that the factory never entirely got right. Mechanical noises were always apparent in the Cub engine, particularly in well used examples. Valve gear rattle, piston slap and timing gear whirrings remained a continual embarrassment to the more sensitive ear but the main cause of complaint was exhaust noise. The later police machines were fitted with a larger than standard

Fig. 5.3c The outer engine cover removed to show the oil filter housing. (Courtesy Alan Halford)

Fig. 5.4a A typical publicity shot. This 1962 machine, T20 82828, is photographed near Meriden. Built in late 1961 it was sold secondhand from the Experimental Dept. to Elite Motors, London in July 1962. (The Mike Estall Collection)

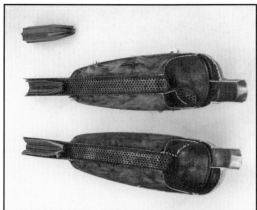

Fig. 5.5 Construction of a 1961 silencer showing the glass fibre packing, deflector and mute. (Courtesy John Nelson)

Fig. 5.4b A dealer's publicity poster from 1962. The models shown are the T20 and the T20SH. Note the cylinder barrel incorrectly painted black. (The Mike Estall Collection)

silencer but the ordinary Cub continued with a series of improvements in this area, none of which was really effective.

Silencer volume was increased several times, in August 1956, August 1957 and again in August 1958, to try and cure the problem. All attempts made a contribution but defeat was finally admitted in March 1959 when an exhaust mute became available. This device, consisting of a rolled and fluted cone, could be fitted to all the current and later machines, but would not fit inside the narrow tailpipes of the earlier silencers. It took the hard

edge off the exhaust note, stifling the flow of exhaust gases using the constriction principle. It worked well enough until the self-tapping screw came loose and the mute fell out onto the road. Then the decibels returned to their former level!

Despite assurances from Meriden, many riders believed that the mute stifled not only the noise but the Cub's performance, and so it was frequently discarded. If ridden with any degree of enthusiasm the Cub was always able to make itself heard and the sublime state of blissful silence was never fully achieved. The Tiger Cub remained a noisy little beast throughout its existence.

5.6 Design changes - carburettor

January 1958 saw the introduction on the roadster T20 of the French-made Zenith carburettor. This was a neat and simple design with initially neither a tickler nor a mixture adjustment screw. The choke size was 17mm, but this was later increased to 18mm, and once the tooling had been made manufacture of the instrument was moved to Britain in mid-1958. At the same time as this move a tickler, a manual cold starting slide and a mixture adjustment screw were added. The slide and needle were now made from brass rather than alloy, as in the original French version.

The French-made instrument was only slightly cheaper than the Amal 332 type it replaced. As soon as manufacture moved to this country customs duty and import charges disappeared and the price came down considerably, falling to 15/- (£0.75), whereas the Amal cost £1.0.9 (£1.04).

Comparative tests had been carried out over a six month period with four machines, using an Amal 332/1 as the standard. It was found that the Zenith gave better acceleration up to 60mph and increased middle range torque. The slow running characteristics were also improved with a better progression from pilot to main jet. However, top speed was marginally worse with the Zenith but fuel consumption was found to be 10% better at 40 and 50mph and 13% better at 60mph.

This Zenith carburettor could be difficult to tune

Fig. 5.4c Top man: 'Airfield boundary in sight, Sir'. Middle man: 'We'll never get this thing off the ground'! Pilot: 'Which one of these is the brake?' (Courtesy EMAP)

Fig. 5.6a Timing side engine detail for a 1958 T20, including the new 17MX Zenith carburettor in use from engine no. 39167. (The Mike Estall Collection)

Fig. 5.6c A restricted Amal 332 for the T20J. (Courtesy James Hallett, Bermuda)

Fig. 5.6b The drive side of the same machine with the new Zenith. This machine, T20 39486, was despatched to Kings of Oxford, Birmingham in January 1958. (The Mike Estall Collection)

and sometimes no amount of fiddling would get a machine running decently. When it performed properly it gave excellent fuel consumption and the engine would run with the machine at any angle of lean, even laying on its side on the ground, whereas the Amal 332 did not like to stray too much from the vertical.

Even as the Zenith was being introduced Amal were working on a replacement for their 332 type, which was intended as a competitor to the Zenith. Amal had already evolved a small monobloc instrument which had comparable performance, but the cost was even greater than for the 332.

The Amal 32 appeared on production T20 export machines in August 1961 and on the home market model from February 1962. This carburettor had been produced experimentally by Amal using a Cub loaned to them in 1958, so there had been plenty of development time. The 32 looked very much like the Zenith but was a much less fickle instrument and became standard equipment on the T20 right through to the end.

The various sports Cubs almost invariably used Monobloc carburettors in two sizes; a 375 type with a

$^{25}/_{32}$in choke for the trials machines, and a 376 type with a $^{15}/_{16}$in choke for the higher performers. One of the American market T20S variants was fitted with a 376 Monobloc that had a remote, rubber-mounted float chamber. The Monobloc carburettor, which needed a manifold placed between itself and the cylinder head due to differing stud centres, was very satisfactory for all types of use and was a good general purpose instrument.

Other types were occasionally used, such as the 20mm Zenith specified for the 'works' trials variant of the T20S, and an Amal R622 Concentric which was used on the last of the Super Cubs. There was also a specialised small choke device for the American T20J 'Junior' model which had either a restrictor in the Amal 332 or a 9.5mm Zenith instrument.

5.7 Design changes - 'works' trials machines

By early 1958 the Cub had gained a good reputation as a more than satisfactory trials machine, and the factory was experimenting with some modifications that would later appear in production models. Here, the combination of low overall weight, adequate ground clearance, reasonable power at low engine speeds and ease of handling, could make even a very average rider look good. Prolonged high engine revolutions were not usual in trials and so the problems associated with that condition were not often apparent.

Not that the bike was perfect: the tendency to run hot was not helped by the crankcase undershield which easily packed a nice layer of heat-insulating mud around the engine. Holes were drilled in the shield and instruc-

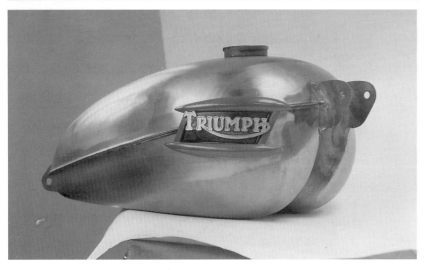

Fig. 5.8a July 1957 and the first trial of the 'Mouth Organ' badge. Note the absence of cross-hatching in the background. (Courtesy John Nelson)

Fig. 5.8b The new one-piece, 'Mouth Organ', badge for the three gallon tank from machine no. 38130. This machine, T20 38254, was sent to W.B. Jury, Tunbridge Wells, Kent, in December 1957. (The Mike Estall Collection)

tions given that riders should run through every puddle they could find to splash some cooling water over the engine! Eventually, the bash-plate was abandoned on 'works' machines, and riders instead fitted tubular rails under the engine.

The factory trials machines were using the slider-block driven oil pump, with its attendant increased flow rate, long before it became generally available. Heavyweight forks from the 350/500cc twins were also used, with slightly shorter stanchions, and fitted with lighter springs. Wide ratio gears, increasing the gap between third and top gears, also came into use. Making its appearance also was a wider swinging fork that was designed for a maximum wheel size of 3.50 x 18in, but into which a 4.00in rear tyre could be squeezed.

The new swinging fork frame could give rise to problems in trials events. Things were far too crowded at the back of the gearbox, there being insufficient clearance between the front of the rear tyre and the rear of the

swinging fork pivot. At the same time, clearance between the left side of the rear tyre, the chain and the chainguard was marginal. Even with the advent of the wider swinging fork, clearance problems remained; to this day anyone riding a Cub in trials must pay particular attention to keeping sufficient space between these components. Failure to keep this area clear of sticks and other rubbish may seriously impede forward progress, even to the extent of bringing the machine to a complete halt.

5.8 Design changes - petrol tank and badge type

Petrol tank design changed a number of times. The first type, used for the Terrier and the 1954/5 Tiger Cub, was 'teardrop' shaped, of $2^5/8$ gallon capacity, and carried the two-piece, 'four bar' tank badge. This tank, with the addition of internal braces, later became the standard fitting on most sports Cubs.

The Cub tank changed to a new 3 gallon design in August 1955 that was much broader than its predecessor, but the Terrier retained the earlier 'teardrop' type. This new 'flat' tank, which was used on the T20 and T20J models to the end of July 1958, carried the two-piece badge until December 1957 when it was superceded by the new one-piece 'mouth organ' type from machine no. 38130.

With the advent of side panels in August 1958 came yet another tank design, this time with a distinct hump at the rear. The 'humped' tank was also of 3 gallon capacity and carried a new one-piece badge. It became the standard T20 fitment right through to the end and was also used on the T20SH sports Cub. An additional feature, brought in during February 1959, was inclusion of internal baffles to reduce fuel surge under acceleration or heavy braking. This tank was also used on the police machines in many instances with four screw holes sunk into the top surface to secure a carrier for radio equipment.

The T20B Bantam and Super Cubs each had their own tank types derived from their Bantam D7 and D10 origins. The Bantam Cub had a modified one-piece

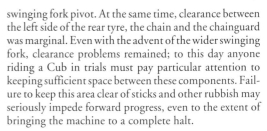

Fig. 5.9a The new 1959 T20 model with side panels and 16in wheels. (The Mike Estall Collection)

'mouth organ' badge and a variation of the two-piece badge was used on the Super Cub.

The two-piece badge consisted of a pressed, chromium-plated styling band over which was placed the cast Triumph nameplate. Both these items were scaled-down versions of the components used on the twin cylinder models and each two-part assembly was secured by two 3BA raised-countersunk screws to a small plate spot welded to the tank side. The styling band used with the two-piece badge came in three different forms depending on which tank it was intended for. The tanks using this badge had different side curvatures at the badge position, requiring styling bands of slightly different shapes. These tanks were the 'teardrop' Terrier and sports Cub type, the 'flat' 1956/58 T20 type, and the one for the T20B Super Cub.

The one-piece badge was also a smaller version of that used on the twin cylinder machines and which became known as the 'mouth organ' type. Each badge was again secured by two 3BA screws but this time into two flush fitting threaded bosses brazed inside the tank. This

badge came in four different varieties, mainly because of differences in the tank side curvatures. The first type was used on the 'flat' 1958 T20 tank and the second appeared on the 1959/60 T20 'humped' tank. The third type, in use from 1961, was the same as the second type except for a cutaway at the front, which allowed room for a chrome styling strip over the petrol tank seam. The fourth type was used on the T20B Bantam Cub.

5.9 Design changes - side panels

In August 1958 semi-rear enclosure was introduced on the roadster T20 reflecting, to some extent, the full rear enclosure on the 350 and 500cc roadster twins. It was considered quite chic by some, though this view was not universal. Rear enclosure certainly made the Cub look very distinctive, but there were drawbacks and the panels were sometimes discarded.

All the side panels machines were fitted with the new 'humped' three gallon petrol tank. At the same time,

Fig. 5.9b A drive side arrangement drawing for a **1960 T20. (The Mike Estall Collection)**

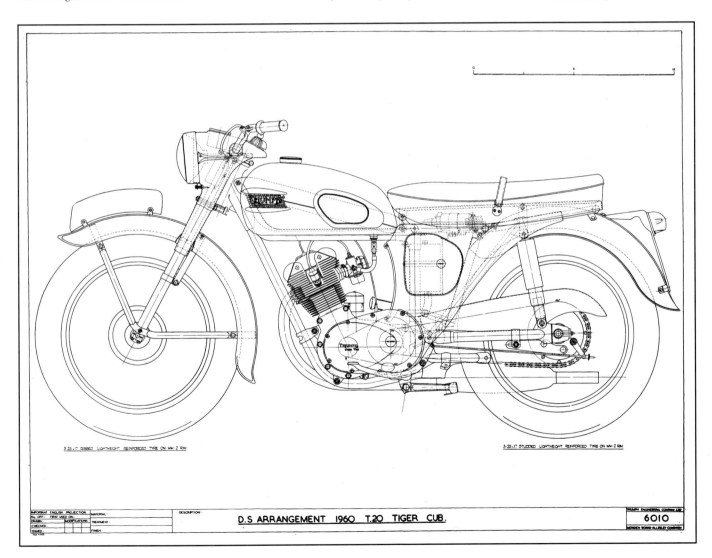

DESCRIPTION: D.S ARRANGEMENT 1960 T.20 TIGER CUB.

6010

Fig. 5.9c The timing side view of a 1960 T20 with the new petrol tank used from August 1958. Substitute 16in wheels and this could be a 1959 T20! (The Mike Estall Collection)

Fig. 5.9d A 1961 T20 restored by the author, in Black and Silver Sheen. A later oil tank with a drain plug was fitted at the owner's request. (The Mike Estall Collection)

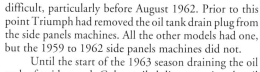

clearance was made between the front of the panels and the carburettor air intake by altering the oil tank and battery box brackets, thus moving these items a few inches rearward.

Rear enclosure was universally disliked by sporting riders on both sides of the Atlantic. Many young bloods of the day considered that it made the machines too cissy-looking and that, even though the Cub was only a lightweight machine, it should still look like a real man's motor cycle. The panels also made routine servicing more

difficult, particularly before August 1962. Prior to this point Triumph had removed the oil tank drain plug from the side panels machines. All the other models had one, but the 1959 to 1962 side panels machines did not.

Until the start of the 1963 season draining the oil tank of a side panels Cub entailed disconnecting the oil pipe junction block under the crankcase and allowing gravity to do its job. To reach the tank properly or to take it off for internal cleaning, meant removal of the right hand side panel.

Removal instructions from Meriden said it was only necessary to undo the front joint of the panels, the nut from the top fixing stud, and the screws at the swinging fork pivot and at the rear of the panels. In reality, congestion around the top fixing stud and the curvature of the panels made it difficult to get a spanner onto the nuts. It was much easier if the seat was removed first, but this was not easy because access to the rear seat bracket bolt heads was obscured by the side panels.

Maintenance requirements sometimes made it necessary to remove the petrol tank. The tank could neither be drawn upwards at the rear because the seat was in the way, nor at the front because it would foul the nacelle, so it was necessary to first remove the seat. Access to both the front and rear seat bolts was awkward because the panels got in the way. They also obscured access to the rear tank bolt.

Careless removal of the panels often resulted in scratches on the paintwork of other components, particularly the oil tank and battery box. A sagging right hand panel could touch the top of the gearbox and engine vibration would cause the panel to wear a groove in the

Fig. 5.9e A 1959 press advert. (Courtesy EMAP)

Fig. 5.9f A 1959 press advert. (Courtesy EMAP)

top of the engine outer and inner covers.

So routine maintenance on a side panels Cub was fraught with difficulty. It was really no wonder that many owners didn't bother to change the oil at anything like the suggested intervals because the panels made the oil tank so awkward to get at. The result was that either the side panels remained in place and dirty oil circulated around the engine, increasing the likelihood of bearing failure or, as was more often the case, the side panels were permanently discarded.

5.10 Design changes - large inlet valve

A new cylinder head with a larger inlet valve came into use from August 1959. This change reflected common practice among engine tuners, making the Cub engine breathe much more effectively, particularly with the 'R' sports camshaft. Some early large valve heads retained the smaller diameter inlet port but soon the larger port head had become standard on all models.

5.11 Design changes - new crankcase and 'two-ball' mains. Revised oilways

From February 1962, at engine no. 84269, a new crankcase came into use with a new timing side main bearing. Originally a stub on the flywheel running in a bush, crankshaft support on this side of the engine now became a ball journal bearing through which ran the flywheel mainshaft. The bush material itself had changed over the years but all the materials wore, as did the cast iron flywheel stub itself, gradually increasing slack within the

Fig. 5.11 The lubrication system from engine no. 84269. (The Mike Estall Collection)

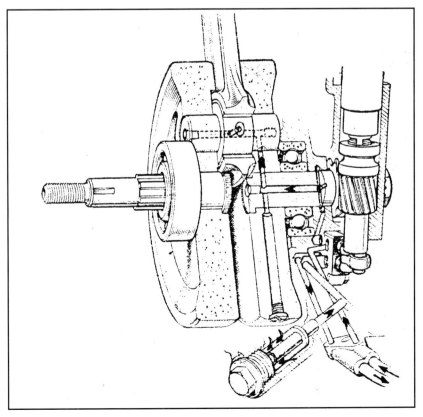

Fig. 5.12 The third engine configuration. Two-piece crankcase, side points, ballrace TS main bearing, 'Oval' head and barrel, duplex primary chain. See also Figs. 3.3b, 3.11 and 5.14. (Courtesy Trevor Gleadall, JR Technical Publications)

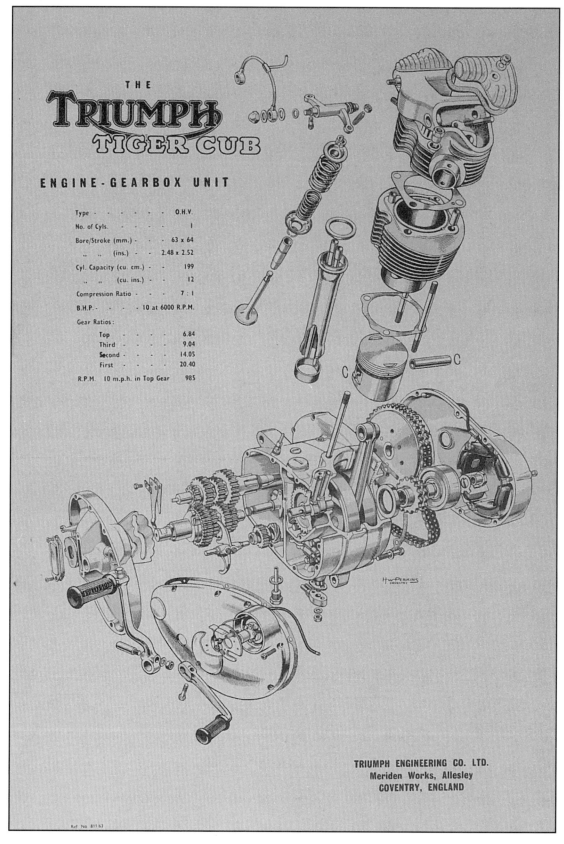

THE

TRIUMPH TIGER CUB

ENGINE-GEARBOX UNIT

Type - - - - -	O.H.V.
No. of Cyls. - - - -	1
Bore/Stroke (mm.) - -	63 x 64
„ „ (ins.) - -	2.48 x 2.52
Cyl. Capacity (cu. cm.) - -	199
„ „ (cu. ins.) - -	12
Compression Ratio - - -	7 : 1
B.H.P. - - - 10 at 6000 R.P.M.	
Gear Ratios :	
Top - - -	6.84
Third - - -	9.04
Second - - -	14.05
First - - -	20.40
R.P.M. 10 m.p.h. in Top Gear	985

TRIUMPH ENGINEERING CO. LTD.
Meriden Works, Allesley
COVENTRY, ENGLAND

Ref. No. 811.63

bearing. As they wore, oil pressure to the big end inevitably declined too. The crankcase oilways were also revised to provide an oil feed to the big end through the new mainshaft.

The new arrangement, in which oil pressure to the big end no longer relied upon the integrity of fit between flywheel stub and main bearing bush, was a great improvement and went a long way towards increasing engine life.

Fig. 5.13 'Coming for a spin?' The 1963 T20 in Flame and Silver Sheen. Note the provision of an oil tank drain plug. (The Mike Estall Collection)

Fig. 5.14 The fourth and final engine configuration. Two-piece crankcase, side points, 'Square' head and barrel, duplex primary chain, no gear indicator. See also Figs. 3.3b, 3.11 and 5.12. (Courtesy Trevor Gleadall, JR Technical Publications)

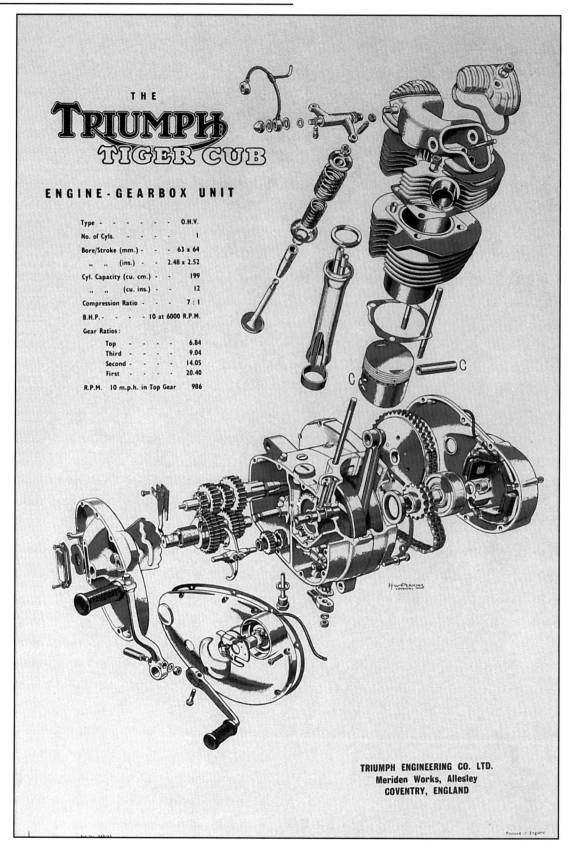

THE
TRIUMPH
TIGER CUB

ENGINE-GEARBOX UNIT

Type	-	-	-	-	-	O.H.V.
No. of Cyls.	-	-	-	-		1
Bore/Stroke (mm.)	-	-	-			63 x 64
" " (ins.)	-	-				2.48 x 2.52
Cyl. Capacity (cu. cm.)	-	-				199
" " (cu. ins.)	-	-				12
Compression Ratio	-	-				7 : 1
B.H.P.	-	-	-			10 at 6000 R.P.M.
Gear Ratios :						
Top	-	-	-			6.84
Third	-	-	-			9.04
Second	-	-	-			14.05
First	-	-	-			20.40
R.P.M. 10 m.p.h. in Top Gear						986

HW PERKINS
COVENTRY

TRIUMPH ENGINEERING CO. LTD.
Meriden Works, Allesley
COVENTRY, ENGLAND

54

5.12 Design changes - new crankcase with 'side points'

In August 1962 the old contact breaker - which perched on top of the crankcase and was sometimes referred to as the distributor - disappeared. A Lucas 4CA points assembly was now located on the end of a new camshaft, protected by an oil seal. The camshaft still retained the original standard or sports profiles, but now provided a location for a new auto-advance unit. As the contact breaker was now driven directly from the crankshaft pinion, it eliminated at a stroke all the mechanical backlash problems associated with the old wear-prone skew gear driven system.

This change did the Energy Transfer ignition set-up much good as well as being tidier in appearance. Location of the auto-advance unit was by means of a peg in the camshaft engaging with a slot in the taper at the back of the unit. This made removal and replacement of the advance unit - necessary if access had to be gained to the oil pump, camshaft or gearbox - possible without losing the ignition setting.

When Cub production moved to Small Heath BSA methods were adopted. They did not use a locating peg for the auto-advance unit and so it was omitted. This retrograde step meant that whenever the timing side of the engine was dismantled, ignition timing was lost and had to be reset. If the engine was an Energy Transfer unit, where correct ignition timing was particularly critical, it could mean that a previous careful, accurate and successful setting would be lost, and might take ages to find again. As the camshaft worked off a half-time pinion, any error in or correction to the physical setting of the advance unit in its taper would be doubled, when related to the piston position, doubling also any change in the ignition timing.

The new crankcase had a redesigned oil collection chamber giving better drainage to the oil pump scavenge pick-up. New inner and outer timing side engine covers also appeared. At the same time the opportunity was taken to provide a rubber-plugged hole in the timing side outer cover through which easy access could be gained to the end of the clutch cable.

The cylinder head now sported new cast alloy finned rocker covers, replacing the chrome-plated steel pressings formerly used. The pressed type usually leaked due to over-tightening of the securing nut, which even doubling the thickness of the cover material had failed to cure, but with the cast alloy type the problem disappeared for good.

5.13 Design changes - twin switches, gear indicator move, new alternators

The side points engine brought about the demise of the complicated and expensive Lucas PRS8 ignition and lighting switch used on the T20 model. In its place came two Lucas 88SA switches, one for ignition control and the other for lighting. In reality these two switches were little more than the old PRS8 split into two parts and provided with moulded plastic plug-in connectors. This arrangement made electrical maintenance much more straightforward as it was easier to disconnect the wiring in the nacelle.

To accommodate the two switches there was a new nacelle top and new, grey-faced speedometer and new cable. This speedometer was also fitted to the sports models, together with a matching tachometer where specified. The gear indicator lost its place of honour on the nacelle top, where it had been since 1953, to become a little indexed finger located on top of the gearbox.

New alternators came into use as well; the RM18 for battery systems and the RM19 for the Energy Transfer machines. The former provided 57 watts at 5000rpm.

5.14 Design changes - cylinder barrel and head finning

One of the Cub's many pitfalls was the tendency to run hot. The finning on the original 'round' cylinder head and barrel turned out, in the light of experience, to be of insufficiently large surface area. In August 1956 a minor improvement to cooling had been made by the addition of some finning to the inlet rocker box on this cylinder head.

A new 'oval' barrel was introduced in August 1958 with elongated finning at the front, giving some additional cooling area. At the same time a new 'oval' cylinder head with further increases in the fin area came into use with this new barrel.

In August 1959, at engine no. 56360, the exhaust port was repositioned by moving it round 7°. This gave a little more room between the exhaust pipe and the frame and allowed the owner to fit the exhaust pipe finned clamp after the engine had been bolted in the frame. Prior to that, if he forgot to fit the clamp to the exhaust stub before putting the engine in the frame, he had to remove the front and lower engine bolts and tilt the engine upwards to fit the clamp.

A further change came in 1960 when shorter barrel studs, coupled with deeper cylinder head nut counterbores and longer nuts, meant that there was now sufficient clearance to slide the head over the studs. It was therefore no longer necessary to lower the front of the engine when removing the head.

The 'oval' barrel and head remained in use until February 1965 when the 'square' type arrived. This barrel had only eight fins whereas the earlier types had nine. Here, the increase in the fin area of both components became really worthwhile and, for the first time, in later combination with the slider-block driven oil pump and the needle roller big end, the Cub engine became as robust as it always could have been.

5.15 Design changes - wheel and tyre sizes, brake drums and speedometer drive

Wheels and tyres came in great variety. The Terrier had 19in wheels shod with 2.75in section tyres throughout its short life. The 1954/5 T20 Cub used the same wheel size but with 3.00in tyres. From 1956 to 1959 for the T20 and T20J the wheel size was reduced to 16in and tyre size increased to 3.25in section: they were universally nicknamed the 'Chubby Cubs', due to their small stature. From 1960 onwards the T20 was fitted with the same section tyre but on 17in rims.

Dunlop Universal or Lightweight block pattern tyres were the norm, front and rear. From 1960 a Dunlop ribbed front tyre was used. Later on (and the date has not been established), an Avon Speedmaster ribbed tyre was fitted on the front and the 'Safety Mileage' SM model used at the rear. Also found on some export machines by 1963 were Dunlop K70 tyres fitted at the rear.

The sports machines and the French Army Cub almost always used a 19in front tyre of 3.00in section and a 3.50in tyre on an 18in rim at the back. The tyre type itself depended on intended use of the model, but were usually Dunlop Universal or Trials Universal. Exceptions to this general rule were the TR20 and TS20 models where 2.75in by 21in tyres, of Dunlop Trials and Sports type respectively, were fitted at the front. At the rear of the TR20 was an 18in by 4.00in Trials Universal and on the TS20 it was a 19in by 3.50in Dunlop Sports tyre. A 4.00in Trials Universal went on the back of the T20SC in 1964/5 and on the Mountain Cubs too.

Yet another change arrived with the T20B Bantam and Super Cubs. Here, tyre and wheel sizes were carried over from their Bantam ancestry: 18in with 3.00in section tyres, front and rear, on both models.

All the plunger framed machines used a speedometer gearbox driving the cable over the top of the rear wheel axle. By August 1956, with arrival of the swinging fork frame, the cable drive was taken from underneath the axle.

The brake drums were always the 5¹/₂in, black painted single-sided type, with the exception of the Bantam Cub where silver paint was used and the Super Cub which had silver painted full-width hubs.

Fig. 5.16a A 'Junior' speedway machine built by Gary Davis from Oregon, USA. It shows just what can be done with a lot of skill and effort. (Courtesy Dan Andrade, Washington State, USA)

5.16 Design changes - pistons, cams and gearsets

Some special engine components, formerly only available on the 'works' machines or developed by tuners in the USA, became more generally available in the late fifties. Items such as a high compression piston, a sports camshaft and alternative internal gearbox ratios were incorporated into the engines of new models or became available as spares.

The T20CA, appearing on the scene in February 1958, was the first of the production models with higher performance parts. Although a high compression piston had been available for some years for the Terrier, the T20CA was the first to be fitted with a 9:1 piston and close ratio gears, but still retaining the standard cam. Another Cub piston, this time giving a 10.5:1 compression ratio, became available in the USA at about the same time but was never fitted to production models.

The sports 'R' camshaft, which was based on the factory 'Q' cam and had the same profile as the famous E3134 cam fitted to the sporting twin cylinder machines, appeared with the arrival of the T20S in December 1958, and was fitted to all three varieties of that model. This cam gave a useful boost in power, becoming basic equipment for most sports Cubs in later years. When used in combination with the 9:1 piston and a large choke carburettor, it turned the Cub into something of a fire-breather!

As well as the close ratio gearbox, the T20S was the first production model to offer wide ratio trials gears or extra-close ratios for racing. The close and wide ratio gearsets each closed or widened the relationship between third and top gears by one tooth, leaving the other gears unchanged. However, the extra-close gearset not only shortened the space between third and top gears by using the close ratio pinions, but also closed up the untidily large gap that lay between second and third gears.

Original selection of the internal gear ratios must necessarily have been something of a compromise. On the one hand the rider had to have a bottom gear low enough to allow only eight or ten horsepower to pull the machine away from standstill, perhaps up a steep incline and while carrying two adults. On the other hand fourth gear had to be such that decent top and cruising speeds were available without overworking the engine.

These two constraints had set the outer limits of the gearing and somehow the other two intermediate gears had to be fitted in between. The final design had settled on gearing which gave a very usable third gear, sitting comfortably below top. The second ratio also sat quite nicely above first but it left a gap between second and third gears that could sometimes be much too wide for comfort.

The extra-close gearset dealt very successfully with that problematical gap and it was fitted to some varieties of the T20S, some T20SS models, the TS20 scrambler and the export only T20SR. One further refinement came when the gap between first and second gears was also closed up by one tooth. This gave an ultra-close box which, although not fitted to any production model, allowed a full racing set-up to be achieved from about

Fig. 5.16b The drive side of Gary Davis' half size speedway bike. (Courtesy Dan Andrade, Washington State, USA)

1960. For racing machines these two gearsets were a great asset. Appendix 3 lists all the factory internal gearsets.

The tooth count on each layshaft pinion, when added to that of its mainshaft companion, always gave a figure of forty-five. Provided this figure was maintained the gearbox pinions could be mixed and matched in many more ways than offered by the five factory gearsets, and produced custom-made ratios to suit the rider's needs. Gears could also be left out altogether, a practice particularly prevalent among American short-track and speedway riders, producing two-speed or even single-speed gearboxes. With a little imagination, tuning the gearbox could be almost as much fun as tuning the engine!

5.17 New models - T20 & T20J

The period from 1957 onwards saw the introduction of several new models. The basic bread and butter roadster T20 remained, of course, but added in the USA from

Fig. 5.17a A 1965 T20 at a stately home. (The Mike Estall Collection)

Fig. 5.17b A different view of the 1965 T20. (The Mike Estall Collection)

Fig. 5.17d & Fig. 5.17e Two
1959 press advertisments.
(Courtesy EMAP)

Fig. 5.17c '... and those nice
people at Triumph call this
bit the engine.' (Courtesy
John Nelson)

August 1957 was the T20J. This model, known as the
'Junior' Cub, was available in certain states where fourteen
to sixteen year old riders were permitted to use low-
powered motor cycles whose engines produced under
5bhp.

The standard T20 Cub was detuned by the simple
addition (and therefore equally simple removal) of a
restrictor plate between the Amal 332 carburettor and the
cylinder head, or by the use of a sleeved 332 carburettor.

This simple expedient reduced the power to 4.95bhp at
5700rpm. From November 1958 a special 9.5mm Zenith carburettor was used, replacing the Amal. In all other
respects the machine was the same as the T20.

The despatch books reveal that only 223 examples
of this model were built and all but a handful went to the
USA, others going to Bermuda, Uganda and one to
England. However, the Tiger Cub and Terrier Register
has discovered other examples in existence that are shown
in the factory records as a plain 'T20', so it's unlikely that
the true number built will never be known.

5.18 New models - T20C and derivatives T20CA and T20CB

Introduced at the Paris Show for the 1957 season was the
T20C, a new model designated the 'Competition Cub'.
The design was greatly influenced by Ken Heanes' gold
medal winning ISDT machine and it was intended as a
road model with the added ability to go off-road with ease
and without modification by the owner. It was the T20C,
together with the new swinging fork chassis, that really
set the competition scene alight in the USA. After its
introduction there followed a quite lengthy period where
this model, and later derivatives, were virtually unbeatable
in all manner of off-road events.

The T20C used the $2^5/_8$ gallon fuel tank and the
new model set some trends that were followed by virtually
all sports Cubs in later years. These included the combination of a 19in wheel at the front with a 3.00in tyre and
an 18in rear wheel with a 3.50in section tyre. Trials or
road tyres could be specified with the result that ground
clearance rose from a lowly $4^1/_2$in on the T20 to a much

Fig. 5.18a A 1957 T20 and a 1958 T20C. Both machines were restored by the author. (The Mike Estall Collection)

Fig. 5.18b An Aztec Red T20C destined for the USA on the Cub assembly track. Photo taken at Meriden by the author 24th September 1957. (The Mike Estall Collection)

Fig. 5.18c A restored T20CA. The front portion of the front mudguard stay is missing and a 1963 type exhaust pipe has been used. (Courtesy Dan Andrade, Washington State, USA)

Fig. 5.19c A scrambler version T20S for the American market. Note the remote float carburettor, tachometer and absence of any lighting. (Courtesy John Nelson)

more useful 6 inches on the T20C. Other new items included an oil tank with a froth tower, a rear sub-frame with a bracket to carry a high level exhaust system, a crankcase undershield and reinforcing brackets at the bottom of the rear suspension units.

Mechanically it was identical to the current T20 except for a smaller gearbox sprocket. With the use of lengthened lightweight forks and the instrument nacelle, it was a very elegant looking machine. The records show that a total of 4111 T20C were sold to 48 countries

worldwide during the 1957 to 1959 seasons.

There was an American version, the T20CA, which had a high compression piston, close ratio gears, a low-level exhaust system and road tyres. Only 365 of this model were sold in 1958 and 1959, all but a very few going Stateside.

Another variant was to have been the T20CB, for

Fig. 5.19a A road specification 1959 T20S, from the timing side. A very good-looker in Ivory and Azure Blue. (The Mike Estall Collection)

Fig. 5.19b The same 1959 T20S, from the drive side. (The Mike Estall Collection)

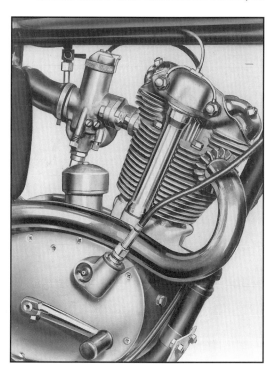

Fig. 5.19d The tachometer drive outer cover. This requires a 4:1 ratio tachometer and a different high-level exhaust pipe whereas the side points engine had a 2:1 tachometer and the normal exhaust pipe. (Courtesy John Nelson)

which an American order of 42 machines had been placed in February 1959. This model would have been exactly the same as the T20CA except for an upswept exhaust system and might have been the replacement for that model. However, the factory despatch records do not show any deliveries of this model and it is assumed that none were sold.

The demise of these three models, the T20C, T20CA and T20CB, came about as a direct result of the introduction of the T20S in December 1958.

5.19 New models - T20S

The T20S (Sports), first seen in December 1958, was another trend-setter and the first of a long line of models suited to either on-road or off-road use. It came in three different guises, all of which had the new features given to the T20C, plus others.

Added now were heavyweight front forks taken from the 350/500cc twin cylinder models, pushing ground clearance up to over eight inches. Each variant was fitted with tyres to match its intended use and all used the new Energy Transfer batteryless ignition system, the new sports 'R' camshaft and a cylinder head with a larger inlet valve. This model was also the first production Cub to use the Monobloc carburettor. These features subsequently became standard catalogue items for nearly all future sports Cubs.

The first T20S machines were the trials variant with

a low compression piston and probably wide ratio gears, then the road and scrambles versions with close or extra-close ratio gears and the high compression piston. The use of a 9:1 high compression piston, sports cam and a larger inlet valve pushed available power up by 45% from

Fig. 5.20a An artist's impression of the roadster version 1961 T20SL looking resplendent in Ruby Red and Silver Sheen. (The Mike Estall Collection)

the roadster T20 model. The scrambles version had a 376 Monobloc carburettor with a remote, rubber-mounted float chamber and usually a 140 main jet, though sometimes a 200 jet would be specified.

Also available were high or low-level exhaust systems. The high-level type had two different bends, for use with the timing side outer cover with or without a

tachometer drive. Another performance enhancement on the scrambler variant was an exhaust tailpipe extension, replacing the silencer. The roadster variant also had lights, speedometer and a tachometer option, too, and it was a real flier. Wheelies were easy on this machine and the factory testers soon found that the bike would comfortably pull well over 60mph climbing Meriden Hill, easily burning-off the 350cc touring twins in the process.

Fig. 5.20c November 1960, the Earls Court Show. Edward Turner and works Financial Director Charles 'Bud' Parker discussing the new T20SL. (Courtesy John Nelson)

5.20 New models - T20S derivatives - model types and specifications

The T20S spawned many new models over the years. They can be divided into three sub-species: those primarily designated as roadsters; those mainly intended for off-road use and those which were 'all-rounders'.

Among the first group were the road versions of the T20S and in 1961 the T20SL (T20S with Lights). Then in 1962 came the T20SH (T20S - for the Home market). The T20SH was unusual among sports Cubs in that it did not rely on Energy Transfer but was fitted with a conventional battery ignition system. For export to the USA from 1962 was the T20SR (Sports Roadster). This model was originally intended for the USA east coast market but a good number went to the Pacific coast salesrooms of Johnson Motors.

There may also have been plans in 1960 to build another road model intended solely for the American market, the T20SA, but no further information has come to light. It is now thought to have perhaps been a model on paper only, having probably been overtaken by development of the road version of the T20S.

The T20S road-going variants were universally nick-named 'Baby Bonnie', not only because they were all hot little road-burners, but also because model colours often echoed those of the 650cc Bonneville twin or its

Fig. 5.20d Another 1960 Earls Court photo. The original caption was: 'A heavyweight on a lightweight!' Henry Cooper (British and Commonwealth Boxing Champion) with Neale 'Sid' Shilton (General Sales Manager). (Courtesy John Nelson)

single carburettor variant, the Trophy.

Off-road models were the trials and scrambles versions of the T20S and 1961 T20T (Trials). There was a trials variant of the 1961 T20SL which, in 1962, was renamed and became the low compression version of the T20SS. Also in 1962 came the TR20 and TS20, which

were trials and scrambles versions of the T20SS. Small numbers of T20W (Woods) models, an economy, off-road version of the T20S, were sold in America in 1960.

The 'all-rounders' arriving in 1962 included the T20SS (Street Scrambler), which was the T20SL road version renamed. Although the T20SS was intended for

Fig. 5.21a A 1965 T20SM Mountain Cub resplendent in Hunting Yellow. (Courtesy Roger Tedds, Ventura, California, USA)

the American west coast, a significant proportion of machines went to the east coast dealers of the Triumph Corporation where it was known as the T20SC (Sports Competition). After that came the Mountain Cubs T20SM and T20M, on sale from 1964 and 1967 respectively. The T20SM was the west coast version of the east coast T20SC. These models were all throwbacks to the original T20C some ten years earlier, in that they were true on/off road machines, being equally at home in either environment.

At the end of 1965 all this model complication

ceased. The whole act was cleaned up and the T20SS, T20SC, T20SR, TR20 and TS20 model designations all disappeared. For the 1966 season only the T20SM and T20SH survived, to which was later added the T20M.

All the T20S derived sports Cub models were variations on a theme as far as their detailed mechanical specifications were concerned. They mostly used Energy Transfer ignition systems, low or high compression pistons, the sports or standard camshaft (mostly sports 'R' type), close, wide or extra-close ratio gears and high or low level exhaust systems. Almost all used the Monobloc carburettor, all had the heavyweight front fork and a crankcase undershield, and most had the 2⅝ gallon fuel tank.

The various combinations of these components resulted in over twenty different model types and subtypes. Also derived from the T20S, by way of the T20T and T20SM, were the military machines, T20WD and T20M.WD, used by the British and French armed forces and the military of a few other countries.

It was as a result of tests made to a sample machine from the forty military T20T models in 1961 that additional strengthening gussets were added to the frame headstock. This feature was adopted on virtually all sports Cubs in later years, starting with the TR20 and TS20 models in March and April 1962.

5.21 New model - the origins of the Mountain Cub

There is an interesting story about this model related by Lindsay Brooke and David Gaylin in their excellent

Fig. 5.21b The Mountain Cub - one of the most successful models ever. This 1967 T20M is all-original except for a reproduction seat cover. (Courtesy Lindsay Brooke, Michigan, USA)

Fig. 5.22a This 1964 machine could be one of several different models: perhaps the USA east coast T20SC or maybe the west coast T20SS or T20SM in Kingfisher Blue or Crystal Blue and Silver Sheen. (The Mike Estall Collection)

book, *Triumph in America*. At a time when Edward Turner was in America and considering winding up production of the many specialist Cub models sold there, three senior American Triumph staff dreamed-up and built an on/off road Tiger Cub. It was designed to compete with Japanese lightweight trail machines then currently in vogue, particularly in the western states. The machine they built was based on the T20SC east coast model, with some elements of the TR20, and had knobbly tyres, wide ratio gears and a high-level exhaust system. They named it the 'Mountain Cub'.

Turner was not too interested at first, but an order for four hundred machines apparently persuaded him that this was a serious endeavour and the first examples of the new model were delivered from Meriden in October 1963. Over the next six months four hundred and seventy five were sent to Johnson Motors in California and twenty to the Triumph Corporation in Baltimore. Had Edward Turner not been persuaded, the world would have been denied one of the most ubiquitous and successful of all the Cub models. The Mountain Cub sold like hot cakes with well over six thousand (or roughly one-third of the total production of all models from 1964) being sent all over the world. Its primary market was the USA, where three thousand went to the west coast and fifteen hundred to eastern dealers. The new model was very popular and American dealers said they could sell every one they could get! The remaining fifteen hundred machines were delivered to twenty four other countries, including over one hundred and fifty to the UK.

Fig. 5.22b The 1965 T20SR in Pacific Blue and Silver Sheen. (Courtesy John Nelson)

Fig. 5.24a Sammy Miller's 200cc 'prototype' two-stroke twin. (Courtesy Jean Caillou, Paris)

5.22 New models with Energy Transfer systems

All sports models from December 1958, with the exception of the T20C, T20CA, T20SH and some later T20SR machines, used Energy Transfer ignition. This was essentially an alternating current magneto, but with its various components spread around the machine, supplying AC current straight to a special ignition coil. Two sets of series connected coils were used in the stator; one set providing direct lighting power and the other for ignition. It was temperamental in use.

5.23 Energy Transfer - operating principles

The principle of the Energy Transfer system was simplicity itself. The alternator supplied unregulated AC to the negative side of the special AC ignition coil. With the contact breaker points closed, this current would be grounded because the ignition coil primary winding was connected in parallel with the contact breaker. But if the points were very briefly opened just as the alternator was producing a peak power pulse, the current would no longer be sent to earth but directed to the coil, which would then produce its spark. The opening period of the contact breaker had to be of very short duration (30°), and accurately timed in order to catch the best current, and therein lay the system's Achilles heel.

A worn contact breaker skew gear produced enough

backlash to make it very difficult for the rider to set the ignition timing with the required precision. Any wear between the contact breaker shaft and body, or in the advance and retard bob weight pivot holes or pivot pins, would make matters even worse. The opening point of the contact breaker and its gap size could vary to such an extent that the necessary ignition timing accuracy was all but impossible to achieve. As these factors accumulated even a very small auto-advance period could be sufficient to cause the ideal points opening period to be missed, thus rendering the system ineffective.

It has also been said that the number of turns of wire in the alternator stator windings were perhaps rather frugal, and that the amount of current therefore generated might have been only barely sufficient for the purpose.

Johnson Motors technicians in America had found that on early production machines the number one engine mainshaft keyway was not properly positioned. This meant that the alternator current could not be in concert with the opening of the contact breaker gap, so that the peak electrical power was not being captured.

Energy Transfer machines, particularly well-used 'distributor' models, would usually either start well enough but misfire at higher engine speeds or, if the timing was re-set to give decent running, would be reluctant to restart if stopped. The factory finally produced a thin walled sleeve that could be slipped over the two distributor spring posts. This limited the movement of the bobweights, reducing the advance range of the unit and enabling both ends of the alternator pulse to be captured

Fig. 5.24b Engine detail from the two-stroke twin. Note the saucepan technology! (Courtesy Jean Caillou, Paris)

within the opening period of the contact breaker.

Some of the Energy Transfer machines sent to the USA were also found to have been incorrectly wired at the factory. The wrong alternator wire had been connected to the stop light switch wire on some T20SR and T20SC models, which resulted in an overload and burnout of the stop light filament. This trouble persisted until Triumph Corporation staff recognised the problem and corrected the fault as the machines were unpacked from their crates.

5.24 Prototype - the two-stroke twin

In the winter of 1956/7 Edward Turner and his design team decided to build an experimental 200cc two-stroke twin cylinder engine. Rumours had been heard at Meriden that a rival firm was going to launch a new 250cc two-stroke twin and Triumph wanted to have something ready if they were true. That rival firm was Ariel Motors, which launched its all-enclosed 'Leader' model in 1958.

The new engine had both barrels cast in one piece, as was the cylinder head. The carburettor fed straight into the crankcase via reed valves and there was a belt primary drive. Much of the design had been copied from the Evinrude Chorehorse as Turner, in co-operation with Bill Johnson of the American west coast dealership Johnson Motors, had also been considering entry into the burgeoning US watersport market. Maybe there was also the thought that it might serve as the starting point for a new range of lightweight machines, perhaps replacing the Tiger Cub?

One of the many problems with the new engine that had to be resolved was the provision of a cover for the Lucas alternator. Close examination of the accompanying photographs will show that the cover used looks rather like a saucepan - exactly what it was! Dennis Austin was sent into the hardware department of a local store in Coventry, clutching a Lucas RM13 stator in one hand and cash in the other, with orders not to return until he found something suitable. He emerged some time later with a saucepan which was an exact fit for the stator. The top and handle were removed and two Cub engine mounting lugs were welded on, the bottom was cut out and the cover fabricated. Problem solved!

The prototype engine was run for some time in a large water tank where it was tested for power output.

Fig. 5.25 The ill-fated ohc twin cylinder engine. (Courtesy John Nelson)

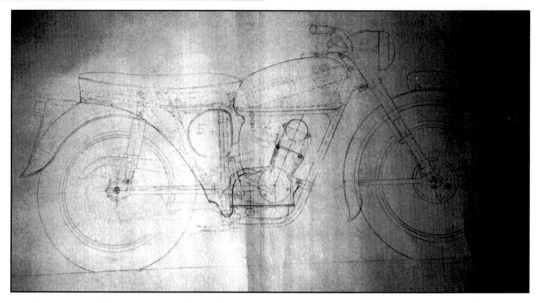

Initially it ran well but, as the test hours built up, performance fell away. The reed valves, exhaust and inlet tract lengths were all changed, but to no avail. It was eventually found that the cylinder liners had been slowly rotating until the inlet and transfer ports were almost cut off. The liners were then pegged into position and the tests recommenced.

Later, the unit was put into a Tiger Cub front frame loop with a Cub gearbox and clutch nailed onto the back, and fresh engine runs carried out outside the factory gates using a mobile test rig. Power output improved but it was then found that the belt primary transmission was using up as much as one-third of the engine's output! A chain was substituted but, by that time, senior management had lost interest and after a short period the project was scrapped, the engine and its vestigial frame being laid to rest in the cellar.

With the closure of the factory came the sale of all the assets, machinery, fixtures, etc. Included in the sale was the experimental 200cc two-stroke engine and frame which were bought by road racer and many times British trials champion, Sammy Miller. He has attached contemporary Tiger Cub cycle parts to the frame and engine and fitted the side panels that would have been standard issue on all T20 machines from 1st August 1958, but modified to suit the absence of an oil tank. Whilst not perhaps the prettiest of models, the resulting machine apparently has performed very well considering it has been built totally 'blind' and devoid of any chance of development time. This most interesting machine can now be viewed in Sammy's museum in Hampshire.

Had the factory persisted with the concept something very worthwhile might well have come out of the experiment. Triumph in general, and Edward Turner in particular, were never big on the two-stroke principle but this little engine might have been just the thing to have set a whole new trend for the Meriden concern.

5.25 Prototype – the overhead cam twin

Reproduced here is a faded and undated drawing from Meriden showing a Tiger Cub with an engine that was clearly something unusual. Cycle parts identify the machine as being from around 1961 but it was the engine that set it apart from other Cubs.

Designed by Edward Turner, the engine was a single overhead cam transverse twin cylinder unit of 200cc. The bore and stroke figures are not known and it looks as if it may possibly have featured wet sump lubrication. The engine was run on the dynamometer but it had small valves, rather inefficient porting, minimal fin area and produced a disappointing 12bhp. It was never fitted into a frame and after a little work the project was abandoned and consigned to the cellar. The unit was sold on closure of the factory but its current whereabouts are unknown to the author.

Chapter 6
Built at BSA

Various models from nos. 101, 2001 & 3001 (February 1965 onwards)

6.1 Move to Small Heath - reasons behind the decision

By the end of 1964 it was apparent that, in order to satisfy the important American market, production of the twin cylinder Triumph models would have to be streamlined and expanded. Following advice from McKinseys, an American firm of management consultants, a decision was taken in July 1964 by Harry Sturgeon, the BSA Motorcycle Division Chief Executive, to relocate the assembly track for the Tiger Cub models from Meriden to the BSA parent company premises at Small Heath, Birmingham, thus leaving Meriden completely free to concentrate on the twin cylinder models. That decision was made public in November 1964 and was not at all popular with the workforce of either factory.

By February 1965 the whole Tiger Cub production line was in full operation in its new location, using engines, frames and cycle parts shipped to Small Heath from Meriden. It was not until August 1965, however, that the first completely Small Heath-built Cubs came off the line. At this point, of course, the frame being used was still the Meriden T20 type. Use of the Bantam frame did not occur until some months later.

Then came the decision to amalgamate all the Triumph and BSA dealerships, the source of further ill feeling between employees of the two factories and dismay of main dealers and the faithful BSA or Triumph rider. Much of the plant and equipment moved to BSA from Meriden was worn out. Crankcase dies, jigs, fixtures, etc., had to be altered or repaired and sometimes packing pieces inserted before they could be used. Problems arose with some crankcases because of the poor condition of these items.

6.2 Move to Small Heath - warranties and build quality

Once relocated to Birmingham, plans were soon being made to do away with the Triumph framed roadster (but leaving the sports models unchanged), and substitute a new hybrid model fitted with a Cub engine but using the Bantam D7 frame and cycle parts. This decision made good economic sense in that it would make the Bantam frame common to virtually the whole of the BSA and Triumph lightweight ranges, but it perhaps did not take into account the strength of feeling among ordinary riders: people who knew what they liked and would stick to it through thick and thin.

Marque loyalty was an emotional concept that Edward Turner had always been at pains to encourage. It was a strange thing, often flying in the face of common sense, but once the virus struck its victim there was seldom any hope of recovery. This concept was now being turned on its head and the rider had to accept a machine that was, so far as he was concerned, neither fish nor fowl. The BSA enthusiast didn't like it because there was a Triumph badge on the tank, and the Triumph man did not feel like paying good money for a bike that had not been made at Meriden.

The Tiger Cub engine and gearbox unit, although now very much more robust and reliable than ever before, was definitely beginning to show its age. It was being put into a set of cycle parts that had changed little in the last ten years and was expected to compete against foreign made, overhead cam single or twin cylinder machines with modern styling and far superior performance.

The sports Cub models in their Meriden frames still remained a force to be reckoned with, but the poor old roadster ... well, it was exactly that! The T20B machines were doomed from the start. If BSA management was pinning its hopes on a revival of lightweight machine fortunes with the two T20B models, then hindsight, with its 20:20 vision, showed how this was not to be.

Another factor also came in to play - again, with detrimental effect. Someone high up in BSA Group management realised that, although the machines were being built at Small Heath, they still had 'Triumph' on the petrol tank. It was therefore considered only right and proper that any warranty work should be undertaken by Meriden and not Small Heath.

The move to Small Heath had not been popular with the workforce of either factory. Anyone at Armoury Road who held a grudge against the Coventry concern (and there were a few), realised he now had a more or less free hand to build into machines being assembled there any mechanical time bombs he wanted, and build quality of the Cub suffered. The consequence was that staff at Meriden were left with no option but to sit back and deal with whatever problems the actions of these malcontents at BSA decided to throw at them. It was all extremely unsatisfactory ...

6.3 Machine numbering - different to Triumph

The numbering system used by BSA was different to that used at Triumph. Briefly, at Meriden there had been one continuous series of numbers, from 101 to 100013, used for all Terrier and Cub models. At Small Heath each model, or sometimes a group of models, was given its own series starting at 101, 2001 or 3001. Full details can be found in Chapter 2.9.

6.4 New model - T20B Bantam Cub

The first of the new hybrid models, announced in September 1965, was the T20B Bantam Cub, sometimes rather confusingly shown in contemporary literature as

the 'T20' model. It consisted of D7 Bantam cycle parts into which had been dropped a Tiger Cub roadster engine. The frame acquired an additional engine mount beneath the crankcase and a pair of small mounting plates added behind the gearbox.

There was a new exhaust pipe, different to the Meriden T20, having a small outward 'kick' by the footrest, and the entry pipe of the silencer was moved outboard of the silencer barrel centreline, instead of being inboard as on the T20. In fact, the silencer body was the same as for the T20 but the bracket had been moved round 180°.

Being now fitted with a four-stroke engine, the Bantam frame had to have an oil tank as well. The wheel sizes were 18in with single-sided brakes and hubs which, although they looked the same as those of the Cub were, in fact, nearly half an inch wider. The Bantam D7 petrol tank was used but with the Triumph 'mouth organ' one-piece badge. The new model was revealed to the public resplendent in Nutley Blue and Alaskan White livery and gold pin-striping on the mudguards.

Once again, in keeping with a long established tradition, there was a delay of several months before the model actually became available for sale in December 1965. It was road tested in April 1966 by *Motor Cycle* magazine and they approved! Performance and economy were very much the same as previous T20 incarnations and the new frame made it feel like a larger machine. It was also competitively priced at just over £165.

In common with all the other 1966 Cub models, the Bantam Cub had the new slider-block operated oil pump drive, giving a 50% increase in throughput from the same sized pump. The Parts List shows that the model came with a caged needle roller big end, but the first few machines still clung to the plain big end that had been used on the T20. The last Bantam Cubs were built in June 1967, although deliveries from warehoused stocks continued long after that.

Fig. 6.4a Artist's impression of the T20B Bantam Cub. (The Mike Estall Collection)

6.5 New model - T20B Super Cub

In November 1966, just over a year after the Bantam Cub was announced, a new 'de Luxe' version was offered - the T20B Super Cub. Literature of the day also referred to this model as 'T20SC' or 'T20', but it should not be confused with the 1962/65 USA sports Cub T20SC, the Meriden T20 model or the T20B Bantam Cub.

The new model, mechanically identical to the Bantam Cub, was road tested by *Motor Cycle* magazine in March 1967. They found, as one might expect, similar performance figures to their Bantam Cub test a year or so earlier.

The Super Cub was a D10 Bantam (later to become the D14), fitted with a Cub engine and having the same frame alterations, exhaust system and oil tank as the D7 based Bantam Cub. The new model came in Bushfire Red, black and chrome-plate and had 18in wheels, this time with full-width hubs, but the same brake linings as on the Bantam Cub. The main colour became known as Firecracker Red in 1968. The Bantam D10 petrol tank used the Triumph 'four bar' two-piece badge.

The Super Cub did not replace the Bantam Cub and the two models ran alongside each other for a year or so with sales of one gradually diminishing as sales of the other increased. The last Cub of any kind to be built was a T20B Super Cub in the week ending 27th June 1969.

6.6 Bantam and Super Cubs - their undeserved unpopularity

Although the Bantam and Super Cubs were considered by many established riders to be unworthy of their affections, representing a betrayal of everything they held dear, in truth, they were good machines, with strong engines and well developed cycle parts. The Cub enthusiast mourned the loss of the Meriden frame and cycle parts that had so successfully projected the charismatic Triumph image. He did not consider the new hybrid machine to be a real Triumph at all, even though it bore the company name.

To the Bantam fan the machine was unacceptable, lacking the simplicity of the two-stroke engine which had been replaced by a more expensive, heavier, clattering four-stroke lump. No, he would stick to the real Bantams, thank you very much!

The unpopularity of the Bantam framed Cubs was borne out by the sales figures. Although sent to more than fifty countries, total sales of all T20B models from December 1965 was 4169 machines, a figure which only just exceeded sales of the T20 model alone in 1964 (3531 machines), the last year before the move to Small Heath and itself the worst year in terms of numbers of Meriden Cubs built.

6.7 The other Small Heath models

Among the other models to be built at Small Heath was the Triumph framed T20 roadster, which was assembled from parts made at Meriden. The last examples of this model may have come off the line in the three weeks

ending 7th March 1966, when the BSA quarterly production summaries show 217 'T20 Standard' models built. However, these machines do not seem to be shown in the despatch records, where the last T20 delivery shown is on 3rd January 1966. This confusion may never be resolved as both the Bantam and Super Cubs were sometimes referred to in Small Heath literature and records as the 'T20'. Without further evidence it is now impossible to say when the last Meriden framed T20 roadsters were actually built.

Also being built at Small Heath was the T20SH sports Cub, which was very smartly dressed in Metallic Blue and Alaskan White, although there were still some in the old Meriden colours of Hi Fi Scarlet and Silver Sheen. The last was built just before Christmas 1965 with subsequent deliveries going through to mid-1966.

The T20SM and T20M Mountain Cubs were being built right up to the first week of April 1968, and the T20M.WD French Army variant was coming off the

Fig. 6.4b Two very pretty models! (The Mike Estall Collection)

Fig. 6.5 An artist's impression of the T20B Super Cub, a D10 Bantam with a Cub engine. The paint pattern on production petrol tanks was not quite as shown here - see Appendix 1.22. (The Mike Estall Collection)

line certainly up to the middle of July 1967 and possibly up to the end of October. Many of these military machines remained in the warehouse for over two years, the final examples not being delivered to Paris until May 1970.

6.8 New model? - the 'Tarbuk Conversion'

This is something of a mystery model, but facts are slowly emerging. In the early months of 1968 forty-six T20B machines were delivered to Elite Motors, Tooting, London. These machines were variously described in the BSA despatch books as 'Tarbuk Conversion', 'Tarbuk Hi Fi', 'T20 Hi Fi', 'SC Hi Fi' or 'Shopsoiled SC Conversion'. The Small Heath build schedules do not show any of these odd descriptions.

Recent enquiries made at Elite Motors revealed a further twenty machines not shown in the BSA records,

Fig. 6.7a An artist's impression of a 1966 T20SH. In 1965 the splined kickstart lever was used and there was no crankcase undershield on production models after August 1963. (The Mike Estall Collection)

200 c.c. TRIUMPH SPORTS CUB (T20S/H)

making a total of sixty-six. Elite's records describe these machines as 'T20B', 'T20SC', 'T20 ex-export', 'T20B SC Clearance', 'T20 Hi Fi Clearance' or 'T20 Hi Fi Tabarrok Clearance'. All these names have been mentioned here as they give some clues about the 'model'.

The Tabarrok Brothers of Teheran were the Iranian Triumph main agent and it is believed that these sixty-six machines were a cancelled export order, originally intended for this agent, but which had been sitting undelivered in the BSA warehouse. Elite Motors was a large customer of Triumph and BSA who sometimes bought cancelled orders or old models and a deal was struck for the whole lot.

The variety of names given to this group of machines is interesting and gives some clues about its specification. The machines are thought to have been T20B Super Cubs fitted with the Bantam Cub petrol tank and badges and possibly Bantam Cub mudguards as well. The use of the word 'Conversion' may have indicated that either they originally had export lighting sets which had to be converted to UK specification, or that they were a converted export order. The 'Tarbuk' appellation is thought to have been a corruption of the Iran agent's name 'Tabarrok', by person or persons unknown in the BSA sales office.

Enquiries at Elite Motors in recent years brought forth a vaguely remembered description of the machines being 'a Tiger Cub engine in a D10 Bantam chassis with the wider Bantam Cub tank and painted in a sort of Dayglo Orange'. A few existing specimens of this machine are now recorded in the Tiger Cub and Terrier Register but unfortunately all have been extensively modified over the years. However, underneath the repainted tinware on a few of them traces have been found of Dayglo Orange paint on a white basecoat!

The best information to hand at the time of writing

is that the machine was a mixture of Bantam and Super Cub components:

1. The D10 Bantam frame was black, as were the chainguard, rear suspension and the D7/D10 type front forks and covers.

2. The fuel tank was the D7 Bantam Cub type with knee grips and a one-piece badge. The word, 'Triumph' on the badge was painted orange or black and the cross-hatched background was either unpainted or in gold. On one existing example the whole petrol tank, oil tank and

Fig. 6.7b The 1966 T20SH in Metallic Blue and Alaskan White, as it appeared in the sales brochure. In production the two colours followed the petrol tank seam and were not as illustrated here. (The Mike Estall Collection)

Fig. 6.7c The author's 1966 T20SH was given the paintwork scheme shown in the sales brochure - the best information available at the time but which is now known to be incorrect. (The Mike Estall Collection)

battery box were all a solid orange colour, sprayed over a white basecoat. In another example the petrol tank had an orange top and silver grey lower half. The orange colour could have been Grenadier Red, the colour used on the late Mountain Cubs, but which for some reason was called 'Hi Fi'. This had no connection with, and should not be confused with the 1964/6 T20 colour 'Hi Fi Scarlet', which was a translucent, flamboyant crimson red.

3. The front mudguard was from the T20B Bantam Cub, centrally ribbed and with chrome-plated pressed stays. It was silver grey and without a painted centre stripe. The rear mudguard was plain section and coloured the same as the front.

4. The wheels were 18in diameter chrome-plated WM1 rims shod with 3.00in Dunlop road tyres. The hubs were the single-sided type from the Bantam Cub, painted silver, or possibly black. The front and rear brake plates were most likely painted the same colour as the hubs.

5. The seat had a grey top over black with dark grey piping and a gold 'Triumph' logo at the back.

6. The headlamp seems to have been the standard Bantam/Super Cub chrome item. It may also have been painted silver grey. It carried Lucas or Wipac ignition and lighting switches.

7. The machine had a low-level T20B exhaust system.

8. The handlebars were the home market, low-level type with bolted on levers and a Wipac 'Ducon' horn and dip switch.

Research continues about this fascinating variation on a Tiger Cub theme and if any reader has new information the author would be very pleased to hear it.

6.9 New model? - T20 Bantam 175

The BSA build schedules show another interesting variant about which little is known. They refer to a model 'T20 Bantam 175', 248 of which were produced between February and May 1969. No details of these machines are known but this model may have been a D10 or D14 Bushman with T20 front forks, petrol tank and mudguards and a 175cc Bantam engine. No such model description appears in the factory despatch books and no other information on this model type has been found by the author.

6.10 Prototype - 'Pastoral Cub'

There was one projected model, sometimes referred to as the 'Pastoral Cub', for use in the Australian outback and in New Zealand that might have done something. The aim had been to produce a machine for under £100, which had a Cub engine in a rigid frame, a platform behind the rider for the sheepdog and a flexible drive for the sheep shearer's clippers, driven from the slot in the old distributor drive shaft. One prototype was built but the BSA board finally decided against it as it was more interested in selling the Bushman model, modified for the same purpose. It finally came to nothing and the prototype was sold. Once again, if a reader has further information or pictures, the author would be pleased to hear of it.

6.11 Total sales from Small Heath

The factory despatch books show 14,265 Cubs of all models delivered from the BSA works at Small Heath during the 1965 to 1968 seasons. This can be compared with just over 108,000 twin cylinder models delivered in the same period from Meriden. (N.B. the 1965 season for Cubs was split between production at Meriden and at Small Heath, but for the sake of example here it has been treated as being all Small Heath).

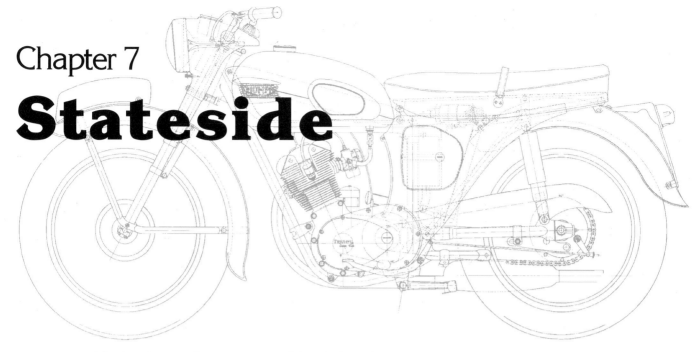

Chapter 7
Stateside

7.1 The American market

The largest and most important export market for all Triumph machines was the United States, and a large proportion of the turnout from the Meriden and Small Heath Tiger Cub tracks headed westwards, across the water to the USA. From the early Terriers in September 1953, through to the last Mountain Cubs in March 1967, the performance and charisma of the Triumph lightweight models was much appreciated 'over there'. Triumph machines had been sold in the USA since well before the first world war so the marque was already well known by the time the first Terriers arrived.

Most machines destined for the USA went by sea and were landed at the port of Baltimore in Maryland. From there they were either sent just up the road to Towson and the Triumph Corporation headquarters, or loaded into railcars for despatch to Johnson Motors, Pasadena, California. However, some of the machines destined for the west coast journeyed instead through the Panama canal to be landed at Long Beach, California.

The factory records show that virtually all USA machines were delivered to one or other of these two giants, and it is most unfortunate that their sales records have not survived. It is, therefore, now impossible to say who the final customer was for any particular machine.

7.2 USA machine specification

USA and UK machines were usually identical except for some of the cycle parts and the colours, but it must also be remembered that the USA had some specialist models that did not grace the shores of the home country.

The very earliest American machines were delivered with UK handlebars, levers and cables but, by late 1953, a 'Western' style high lift handlebar had become the norm for the American market. On the nacelle models these bars were fitted with bolted-on folded type levers instead of the solid levers fitted to the UK and general export market handlebar, which had brazed-on lever brackets. Some USA machines also had a safety strap across the middle of the dualseat for the pillion passenger to hang on to. Full colour scheme details for all models are shown in Appendix 1 and are summarized in Appendix 5.

7.3 Triumph Corporation - 'TriCor'

The motor cycle market for Triumph in the United States had been divided up into two regions. The east

Fig. 7.2 A student mechanic working on an early Terrier engine at the Triumph Corporation Service School. (The Mike Estall Collection)

Fig. 7.5 A Tiger Cub decal from the USA, artist unknown. (The Mike Estall Collection)

Fig. 7.6a The new USA 1954 T20 outside the CBS offices in Los Angeles. The other 'model' is TV news and sports writer Gil Stratton. (Courtesy R.E. Rogers, Little Rock, Arkansas, USA)

owned subsidiary of Triumph in England and controlled sales in the eastern two-thirds of the USA.

7.4 Johnson Motors - 'JoMo'

The western states, with roughly one-third of the population and land area, were looked after by an independent business founded by Bill Johnson, called Johnson Motors Inc. This business had been set up in 1938 in Los Angeles to sell both American and British machines, but, as the years went by, it had become increasingly devoted to Triumph models. By 1965 this business, too, had come under the direct control of BSA, just like its eastern counterpart.

7.5 Sales to the USA

The rate of delivery of the first Terriers to this most important market was disappointing at the start, so much so that the Triumph Corporation had to write to all its roughly two hundred dealerships in October 1953 apologising for the delay and hoping that things would pick up by December - and so they did.

The first Terrier, a pre-production machine, had arrived in the USA in February 1953. It was just a sample of the merchandise, intended to whet the appetite, but shipments of production Terriers were slow at the start. Only sixteen machines were sent in September 1953; eight to JoMo and eight to TriCor. In October forty machines were sent, followed by fifty in November, but then about two hundred went over in December, followed by over three hundred in January 1954.

coast and many central states were the province of the Triumph Corporation of America, which had been founded by Denis McCormack in 1951 and which was based in Baltimore, Maryland. This business was a wholly-

Fig. 7.6b Another view of the new 1954 USA T20 model. (Courtesy R.E. Rogers, Little Rock, Arkansas, USA)

7.6 Early machines rationed

The Triumph Corporation instructed its dealers to sell only one machine of the first sample pair received - and then only to a good customer who could be relied upon to parade the machine around town to encourage further interest. The second machine had to be kept unsold and in absolutely pristine condition to be either ridden around by the dealer himself or kept for showroom display.

7.7 Export packing

Export machines for most markets were usually packed, (like kippers) two in a box with the front forks, front mudguard, seat, exhaust system, wheels, footrests, rear number plate, kickstart and gearchange levers removed and stored elsewhere in the wooden packing case. As an alternative, and also for delivery to the home countries, they would be fully assembled but wrapped in corrugated cardboard tied on with string. In later years, if there were large numbers of machines going to the same export destination, they would be packed sixty at a time into containers.

When received in the USA the machines had to be reassembled and checked over before they were fit for sale. The battery had to be filled with acid, all nuts and bolts checked for tightness, all necessary lubricants loaded and swinging fork pivots greased, tyre pressures checked, valve gear, chain and brake adjustments made.

It was recommended that, where appropriate, the carburettor needle should be raised one notch to provide a richer mixture for cooler running before the five hundred mile first service. The new machine was then road tested for several miles and all items rechecked before being put in the showroom.

7.8 Announcement of the Tiger Cub in the USA

The Triumph Corporation missive containing these instructions (dated 29th October 1953) also gave notice that the factory was working on a 200cc model, a sports or 'Tiger' version of the Terrier, to be known as the 'Tiger Cub'. In the light of recent experience it was made clear that orders for the new model would not be taken just yet but only after a proper period had elapsed, allowing time for development to move past the purely experimental stages. Meriden's fingers had been burnt with the delay and problems in getting the Terrier into production and they did not wish to repeat the experience with the Tiger Cub.

Tiger Cub deliveries to the USA were a little more prompt than had been those of the Terrier, and one machine was delivered to Johnson Motors in March 1954, followed in April by fifty-one more with forty-one going to TriCor. Thereafter, well over a hundred were delivered in each of the following few months. Then it was approximately one to two thousand motor cycles per annum, peaking in 1960 with 2195 machines, more than half of which were T20S models. Altogether 21,453 Terriers and Cubs, or nearly 20% of total production, are shown in the factory despatch books as going to the USA.

In the beginning the Terrier retailed at $449 and the Cub at $499. With an exchange rate then of approximately $4 to £1, this was very roughly the same price as in the UK, including Purchase Tax.

7.9 Dedicated USA models

The American market had been important enough to generate development of dedicated models specifically

Fig. 7.8b A 1956 USA advert. There was nothing 'all-new' for the Terrier, except reversion to the Amaranth Red paintwork! (The Mike Estall Collection)

designed for the conditions found there. The T20CA, T20J, T20SS, T20SR, T20SC, T20SM and T20M were all designed and built with the USA in mind and the genesis of these models has already been described.

In the UK the sports model engines were reported as delivering 14.5bhp at 6500rpm, but the USA preferred a slightly higher figure, often 15.1 or 16bhp. One might have been forgiven for thinking that this was pure showmanship, but it was more likely to have been due to regulations imposed on vehicles using certain types of road. In California, for example, highway regulations on the multi-lane freeways forbade their use by pedestrians, cyclists and motor cycles under 15bhp.

7.10 Colour schemes

USA machine colours were usually a year ahead of the home and general export models, but there were also one or two liveries more or less specific to the States. For example, between 1958 and 1960, several models were available there in Aztec Red and black, a rather garish orange-red, which was definitely not to the rather more conservative tastes of the British market at that time! However, this colour scheme was not exclusive to the USA and a few Aztec Red machines were delivered to some European countries, including one to the UK, and to other regions in the Pacific basin.

The first of the alternative colour schemes arrived in 1955 when the American market Terrier, hitherto resplendent in Amaranth Red, suddenly appeared in black. This only lasted one year and the colour scheme reverted to 'Ammer and Thread' for the rest of this model's short life.

The Mountain Cub in 1965 was offered in Hunting Yellow. This colour was said to have come about in an attempt to dissuade trigger-happy hunters who roamed the Californian mountains from taking involuntary potshots at passing trail riders, by making the machine more conspicuous. This is perhaps an unlikely tale but the author has it on good authority that it's true. The same colour also infected the T20SC for that year but lasted for one season only, the Mountain Cub becoming Grenadier Red and Alaskan White and the T20SC model not surviving into 1966.

It should be noted that USA colours were also sometimes the same as in other markets but were given different names. For example, the USA's Flamboyant Ruby Red was known in the UK as Burgundy, which later became Hi Fi Scarlet. The UK's Kingfisher Blue was known over there as Pacific Blue. A few examples of machines with these USA colours did leak out but they were mostly confined within American borders. One other American touch on the T20 roadster was that the front brake plates were usually chromium-plated, a luxury denied the home market!

7.11 After-market performance parts

In the USA, and particularly in the competitive hotbed of the western states, performance was everything. Even by

Fig. 7.10a July 1959 and three visitors from the USA collect their Aztec Red Cubs from Meriden. Anne and John Adden from Boston and Sue Mason from New York subsequently toured much of Europe and Scandinavia before returning to Meriden for their bikes to be shipped to America. (Courtesy John Nelson)

1954 a kit, more fully detailed in Chapter 9, was available in the USA to uprate the power output of the Terrier by as much as 80%. By porting the head and using a larger inlet valve, fitting the new Californian-made 'Big Bear' cam, a high compression piston, stronger valve springs and regrinding the tappet radius, top speeds could be pushed to near 90mph. Special fuel mixtures were also used for speed records such as those on the California dry lakes and at the Bonneville Salt Flats.

Swinging fork conversions were available for plunger framed machines, as was a lightweight rigid frame for 'short-track' events, and Earles type leading link front

Fig. 7.10b An artist's impression of the 1961 T20 for the USA. (The Mike Estall Collection)

Fig. 7.10c An artist's impression of the 1963/64 JoMo T20. (The Mike Estall Collection)

forks from Californian, Bill Martin. The competition rider could also purchase a heavy duty clutch kit, Sparks and Witham racing valve springs and a range of gearbox and rear wheel sprockets. On offer too were sports

Fig. 7.11 The legend on the back of this picture reads, 'Outside the Hampton Blvd. store, Norfolk, (Virginia), Jan. 1955. Lots of love, Dave'. The rider is believed to be David Jones, ex-Meriden Experimental Dept. and TriCor. He was also Jack Wickes' brother-in-law. (The Mike Estall Collection)

camshafts, a Bates flywheel magneto and a bolt-on oil filter kit with replaceable cartridges. Another ploy used by the more inventive tuner, before the arrival of sports valve springs, was to use Ariel Square Four outer springs together with Triumph 3T inners.

By 1960 the American enthusiast could buy a 250cc conversion that both bored and stroked the engine and used a very high compression piston. A modified flywheel assembly allowed the use of the latest large diameter big end bearing. He could get hold of tubular alloy pushrods, roller-ended cam followers, special piston rings and valve springs, alloy valve collets and an offset rocker ball pin which increased valve lift. For all-out speed there was also much use of special fuels, oversize carburettors and single speed gearboxes with dry clutches.

Available from companies such as Webco of Alhambra, California, were cast alloy rocker covers, with or without breather tubes, a cast alloy, finned rocker oil feed banjo and an oil pump with a pressure relief valve, plus much more. By 1962 an ultra-close gearset was available, narrowing the overall internal gearing of the extra-close gearbox from 2.27:1, making it 2.05:1.

7.12 Competition in the States

Terriers and Cubs had been used in all sorts of events right from the earliest imports in 1953 and 1954, but it was the arrival of the T20C with its swinging fork frame that really ignited their success in competition. Here at last was a chassis that could make good use of the already available engine tuning parts. The Meriden lightweights then performed with great distinction until they became outclassed by the arrival of more modern machinery in the early sixties.

Fig. 7.12 A typical club or AMA paddock scene. The machine is a remote float T20S. The rider and his companion are unidentified. (Courtesy Lindsay Brooke, Michigan, USA)

There were also some events specifically intended for lady riders for whom the Cub, with its ease of handling, would have been ideally suited. These events were often, somewhat patronizingly, known as 'Powder Puff Races'.

7.13 Types of events in the USA

See accompanying table (right).

7.14 Major Enduro successes

The Catalina Grand Prix was over 60 miles in length and held on Catalina Island off the coast of California. In the capable hands of such riders as Don Hawley, Hazen Bair, Ed Kretz Jr., John Smith and Ralph Adams, Terriers and Cubs won 165cc and 200cc classes six times in the four year period 1955 to 1958. This event also attracted riders more famous in cinema such as Lee Marvin and Keenan Wynn, who no doubt enjoyed themselves hugely when away from the film set. They were both rated as top-notch amateurs but sometimes failed to finish! Another top rate rider from the world of Hollywood more usually associated with the large capacity twins was Steve McQueen. He, too, sometimes rode a Tiger Cub.

The Big Bear Run took place over 150 miles of desert terrain. The 165cc class was won by Bobby Skibstead in 1955 and the 200cc event by Wilbur Lamoreaux in the same year and by Kenny Harriman two years later. Triumph riders took the honours in the 165cc and 200cc classes in 1959.

The Jack Pine Enduro was run in Michigan and Cub rider John Wright took the lightweight class and overall honours in 1962. California's fearsome 530 mile Greenhorn Enduro fell to Cub riders Frank Heacox in 1957, Gene Retroff in 1959 and the famous Bud Ekins in

Description	Track type & names	Lap or length	Eligible machines
Flat track or Dirt track	Dirt ovals - often horse racing tracks	$\frac{1}{2}$ to 1 mile	Big bikes - 500cc Not Cubs
Short track	Indoor or outdoor - paved, cement or wood. Santa Fe Park, Illinois. Long Beach arena, San Diego, Fresno and Sacramento, California. Cincinnati Gardens	$\frac{1}{10}$ or $\frac{1}{4}$ mile	250cc machines Rigid framed Cubs best until arrival of Spanish 2-strokes and Harley/Aermacchi 4-strokes.
Speedway	Cinder track - as UK	$\frac{1}{4}$ or $\frac{1}{2}$ mile	500cc JAPs, ESO's. Not Cubs
Junior Speedway	As Speedway	$\frac{1}{4}$ mile	Machines up to 250cc
Enduro	ISDT type events - mud, water, desert etc 'Jack Pine', Michigan. 'Catalina', California 'Big Bear', California 'Greenhorn', California	100s of miles. 500 miles 60 miles 150 miles 500+ miles	All classes Cubs dominated Lightweight events
Road racing or Short circuit	Road or tarmac track	Lap lengths varied	All classes. Cub best in Lightweight classes until Italian 250s arrived
TT, Scrambles, Steeplechase	Twisty, dirt surface with at least one jump	Lap usually below 1 mile	All classes
Trials	Rough ground	Short 'sections'	All classes
Hill climbing	Very steep hills with loose surfaces	Varying length of climb	All classes
Sprinting or Drag-racing	Dead straight track	Varying lengths	All classes

Fig. 7.14a Hollywood stars Lee Marvin and Keenan Wynn. Both were competent Cub racers in the fifties. (Courtesy R.E. Rogers, Little Rock, Arkansas, USA)

Fig. 7.14b This beautifully restored 1960 scrambles T20S belongs to Ray Corlew, Tennesee. (Courtesy Dan Andrade, Washington State, USA)

members carried out all the construction and maintenance required using plant and machinery loaned by nearby sympathetic heavy equipment operators. Every weekend in the fifties and sixties hotted-up Tiger Cubs, particularly the T20S model, were thrashed around the many tracks, often giving riders who later became famous their first competitive rides.

It was not unusual in this period for Cub mounted riders to fill the first several places in some events, culminating perhaps with a ten Cub whitewash in the April 1960 Tecate Grand Prix! At that time there wasn't much that could keep up with a good pilot on board a T20S Cub in the 200cc class.

7.15 Cubs on 'short tracks'

Short track events were a Cub playground for all classes of riders with experts, amateurs and novices often all riding together in the same race. Usually on tenth-mile tracks, these races were short, sharp and often violent. Commonly of five or ten laps only, they were ideally suited to the Cub with its excellent power-to-weight ratio. But on the longer quarter-mile ovals and with the arrival of the Bultaco two-strokes and Harley/Aermacchi four-stroke singles, sheer horsepower held sway and the Cub struggled.

The Cubs were, of course, highly tuned and often had custom-made, weight-saving rigid frames converted from plunger models. On a one-tenth mile of mirror-flat

1964. (It was Bud Ekins who did the barbed wire fence jump for Steve McQueen in the Hollywood film *The Great Escape*. McQueen, himself an accomplished rider, would have done it but for insurance constraints).

A particularly popular form of racing was on scrambles or TT tracks, especially in California. Many of these tracks were sponsored by local bike clubs whose volunteer

Fig. 7.15a Frantic action was normal in short-tracking! The riders are unknown but the venue is the Long Beach track, California, USA. (Courtesy Dan Mahony)

Fig. 7.15b More action from Long Beach. Note the steel 'shoe' on each rider's left foot. Ken Maely has been making these shoes for the racing fraternity for over forty years. See also Fig. 9.10. (Courtesy Dan Mahony)

boarded or concrete track there was little point in having any suspension at the rear anyway! There was only one capacity class and many riders fitted proprietary 250cc kits, some of which were later approved by Meriden and put into the JoMo and TriCor catalogues. Many riders ran two-speed gearboxes by removing some of the pinions, increasing lightness by drilling unwanted metal from every possible component, and used any trick piston and cam they could lay their hands on.

Short track specialist Eddie Mulder was virtually unbeatable at the California Long Beach tenth-mile indoor arena, and only a very few top riders like Butch Corder could make any impression on his string of successes there. Gary Nixon took three national titles on a Cub at the Santa Fe Park, Illinois, quarter-mile track in 1963, 1965 and 1967, and Cubs won huge numbers of other minor events all over the place and for many years.

Gary Nixon, as well as being a rider of the highest quality, was something of a showman who liked to 'psyche' the opposition whenever possible. One of his most successful short track mounts was a 250cc Cub built by Dick Bender and Cliff Guild of TriCor. This machine had a rigid rear end with the wheel axle mounted on a pair of alloy plates that could be quickly unbolted, allowing the wheel to be rapidly changed or the plates themselves changed, easily altering steering geometry. Having no brakes, only the chain had to be broken and the bolts undone to make the change. The machine also employed reinforced rubber tubes to drain the oil from the rocker cavities into the pushrod tunnel, as the normal drain

passages in the barrel had been removed by the boring and resleeving process used to achieve 250cc.

This particular machine was equipped with Nixon's special 'flame-thrower' exhaust. This fiendish device consisted of a spark plug mounted at the very end of the exhaust megaphone and connected to the HT coil.

Fig, 7.15c Did he fall or was he pushed? (Courtesy Dan Mahony)

83

Fig. 7.15f Gary Nixon with his 250cc short-tracker at Daytona in 1995. Here seen pointing out the 'trick' flame-thrower exhaust. (Courtesy Lindsay Brooke, Michigan, USA)

When bump-starting the bike, Nixon would hold the 'kill' button down until sufficient unburnt charge was loaded in the megaphone. On releasing the button, the spark would light the charge with a resounding report and a spurt of flame rearward. It was not unknown for this contrivance to be used during a race to upset any opponent near enough to be affected. Nixon also let slip that it made the bike go faster, too, so it wasn't long before other machines appeared fitted with this particular tuning aid.

7.16 Other events

Road races at many tracks - including Daytona, Watkins Glen and Laconia - were won by Triumph lightweight

Fig. 7.15e This plunger T20 frame, T8857, has been converted for short-tracking. (Courtesy Jonathan Brown, Carpinteria, California, USA)

Fig. 7.15g Engine detail of Gary Nixon's 250cc machine. Note the rocker box oil drains. The machine was built by Dick Bender and Cliff Guild at TriCor. (Courtesy Lindsay Brooke, Michigan, USA)

riders for the best part of a decade from 1954. One such rider was TriCor chief mechanic Cliff Guild who also gained a tremendous reputation for building extremely fast Cubs. Ed Fisher from Pensylvania won the lightweight Daytona event in both 1959 and 1960 on a Cub. Other tuners too, like Jack Wilson and Danny Macias, built drag-racing Cubs capable of exceeding 'the Ton' on nitromethane and other forms of liquid dynamite.

Texan Jack Wilson's Cub dragster engine was fitted into Johnnie Allen's world speed record shell in 1957, powering the projectile to 136mph. This engine used, of all things, the very early Terrier roller big end - the same type that Meriden had so quickly replaced by a plain bearing because reliability had been suspect. Wilson had found by experience that this bearing was the best for his particular type of application.

Terriers and Cubs were campaigned with some success in many other short circuit road race events. With a large choice of tuning hardware, the Triumph lightweight rider did very well until the Ducati/Parilla single cylinder four-strokes and later Yamaha two-stroke twins appeared on the scene. These were altogether more modern designs and the two-strokes in particular had benefited hugely from European research into exhaust and induction tract design. So, by the mid-sixties, the Cub had become completely outclassed and was no longer seriously competitive in this type of event.

7.17 A world speed record

Perhaps the final accolade should go to Bill Martin, a 49 year old Triumph dealer from Burbank, California. He constructed and rode a petrol-fuelled streamliner on the Bonneville Salt Flats, Utah, on 26th August 1959, collecting an AMA 200cc Class C one mile record. He produced an average two-way run of 139.82mph, including a best one-way speed of very nearly 150mph.

The engine was taken to 9100rpm and used standard valves, a 10:1 Robbins piston and 'R' camshaft, but it had been ported and used lightened rockers with Johnson Motors' own S & W outer valve springs and Webco alloy pushrods. An Energy Transfer ignition system was used and a 1^7/$_{32}$in Amal GP carburettor with a remote float chamber. The inlet and exhaust tract lengths had been carefully researched from a book called *The Sports Car* by Colin Campbell, and the exhaust system was a straight pipe without megaphone. The streamlined body of the record breaker was made from a converted P38 Lockheed Lightning aircraft 150 US gallon drop tank.

The drop tank shell was made in three sections, each part being bolted to the next through internal rib formers. No part of a Tiger Cub frame was employed but two quarter-inch thick alloy plates were used as bulkheads, the various engine and cycle parts being secured to them.

Shortened Cub lightweight fork stanchions were used and the sixteen inch wheels were shod with Metzler tyres. The whole front end was bolted to the forward bulkhead and steering was achieved by means of a drag

Fig. 7.15h Rear end detail of Nixon's machine showing the quick-change rear wheel and 'flame-thrower' exhaust system. (Courtesy Lindsay Brooke, Michigan, USA)

Fig. 7.15i Don Getz, a district champion, in action at Paradise Mesa Raceway in 1958 or 1959. (The Gordon Menzie Collection)

Fig. 7.15j Many times winner on Cubs 'Skip' van Leeuwen. The event shown here is probably the Grand Prix on Catalina Island, California. (Courtesy Dan Mahony)

link from a short handlebar in the cockpit. The turning circle was about one hundred feet. The rear end was without any form of suspension and consisted of an axle carrier and wheel bolted to the aft bulkhead by a short section of fabricated framework.

Even though there was no roll cage the idea was that at the start of each run the machine was held upright in a wheeled trolley, but at the end of the run was designed just to fall over on its side. The machine was found to be stable down to about 10mph, so this method of arrival was more undignified than dangerous.

The machine could be difficult to control in anything more than almost calm air. During the record breaking runs there had been a side wind of about 12 to 15mph, causing the whole machine to lean slightly to one side or the other, just like a yacht. It was found that the slight list to port or starboard changed the level of fuel in the float chamber, causing carburation problems that, due to time constraints, could not be cured. It meant that the engine was running lean in one direction and rich in the other. This is the reason why there was such a large disparity between the speeds in each direction - almost 150mph one way and just over 130mph the other.

Nevertheless, an average speed of almost 140mph was attained. This was quite an achievement and showed what could be done with courage, commitment and the careful choice and assembly of 'off the shelf' components.

Fig. 7.15k 'Skip' van Leeuwen in action again. (Courtesy Dan Mahony)

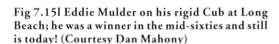

Fig 7.15l Eddie Mulder on his rigid Cub at Long Beach; he was a winner in the mid-sixties and still is today! (Courtesy Dan Mahony)

Fig. 7.17a Bill Martin, (left), his two sons, Dale and Lennie, and the record breaker outside Johnson Motors shop in Pasadena. (Courtesy of Dale Martin, Burbank, California, USA)

Fig. 7.17b Bill Martin setting out for a high speed run at Bonneville. (Courtesy of Dale Martin, Burbank, California, USA)

Chapter 8
The 'Bermuda Cub'

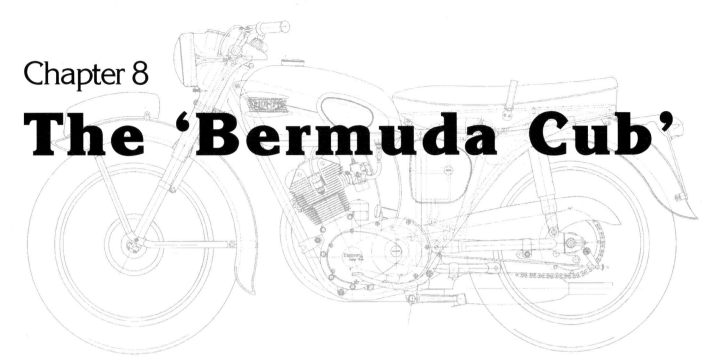

8.1 Capacity and bore limits

When Edward Turner visited Bermuda in 1951, the use of powered two-wheeled vehicles on its roads had only just been allowed and, among the many regulations then in force, was a maximum 2.1 horse power limit. This figure had no connection with the engine's brake horse power but was based on a formula which equated bore size with engine power. Later on, in 1956, the horse power formula was abandoned in favour of a straight 150cc capacity limit. The Terrier was within bounds but this change also coincided with the importation of the first Tiger Cubs and they had to be fitted with 150cc engines in order to comply.

The 1956 legislation also corrected a basic flaw in the current regulations, earlier realized by the island's governing body: it had omitted to limit the flywheel stroke or the number of cylinders an engine could have, so that, initially, the regulations were self-defeating. As a result there had been, in reality, no limit to engine size at all. Therefore, the new capacity limit was set at a figure which coincided with the largest engine sizes already on the island which were Terriers and various 150cc Villiers powered machines.

It is interesting to speculate that perhaps a seed might have been sown at the time of Turner's 1951 visit. It may also have been something more than pure coincidence that the bore and capacity figures he finally settled on for the Terrier were so similar to those permitted by the island's regulations. Bermuda was a ready-made export market if ever there was one, with a populace just itching to get its hands on any form of two-wheeled transport!

8.2 Special model, numbers delivered

Bermuda was unusual in that it was one of the few markets (the USA and France were the others) where special models were produced with features peculiar to that market's requirements.

This island, about six hundred miles off the east coast of the USA, is approximately twenty-two miles long and between one and two miles wide. With a population after the war of around 45,000 - and of strongly British heritage - Bermuda had very stringent regulations on the use of motor vehicles and all machines had to conform. Between the years 1954 and 1964, a total of 310 Terriers and 737 Cubs were delivered there, representing just over 1% of total Cub and Terrier production during those years, so this was a significant market for Triumph.

To understand the many regulations in force concerning motor vehicles in Bermuda, it may be helpful to give some historical background and detail.

8.3 Legislation from 1905 to 1947

The first legislation involving motor vehicles was enacted by the island's House of Assembly in 1905, followed by more two years later. In 1908, further legislation came into being banning the use of all private vehicles as the roads were considered unsuitable. The few vehicle owners of the time were compensated for the confiscation of their conveyances. The only private vehicles subsequently permitted on the island's roads were bicycles and horse drawn carriages, except some lorries for limited commercial use.

In 1942 the 1908 Act was temporarily amended to allow the use of military vehicles during the war, but still no private vehicles were allowed. This wartime temporary legislation was allowed to expire on 31st December 1945 but it was not until early 1947 that legislation was finally enacted authorising, for the first time since 1905, the use of private motor vehicles on the island's roads. However, motor cycles were specifically disallowed.

In February 1946, an Act had been passed permitting the temporary use of 'auto-bicycles' (not motor cycles) for testing and demonstration purposes. Despite the influence of a very strong anti-motor cycling lobby, the 1946 Act had allowed the public to decide for itself whether the use of these vehicles was considered desirable.

Even though private vehicles were now being given some limited use, there were still strong feeling amongst some of the island's more influential characters that they shouldn't be allowed into private hands at all! The newly passed legislation that brought the island into the Twentieth century was therefore liberally sprinkled with conditions restricting the form and use of private vehicles.

8.4 1946 legislation, definitions

The February 1946 Act defined an auto-bicycle as:
(a) not a Motor Cycle.
(b) a motor assisted pedal bicycle -
(i) the wheels and frame of which were of a design and construction similar to a pedal bicycle, and which could be propelled in the same manner as a pedal bicycle.
(ii) the motor should not exceed $1\frac{1}{2}$ horse power.

Another requirement was that the cylinder bore size should be from 40.3mm to no more than 58.19mm, depending on a horse power classification. The class which later came to include the motor cycle as we know it was that of 2.1hp.

The horse power formula was calculated by squaring the cylinder bore dimension in inches and dividing the resulting figure by 2.5. Thus, a bore of 1.58in worked out at 1.58 x 1.58 ÷ 2.5 = 1hp. The maximum permitted bore for the 2.1hp class worked out at 58.19mm.

During his 1951 visit Turner would have been informed about the need for operable pedals, and a machine so constructed and designed that it could be operated in the same manner as a pedal bicycle. It has been said that the first Terrier to arrive in the island (date unknown, but prior to delivery of the first production machine to go there in September 1953, and also presumably prior to the March 1953 legislation which removed the pedals requirement) is reported to have had pedals mounted on some sort of bracket, allowing the machine to be propelled like a bicycle.

It is most unfortunate but no pictures of this contrivance exist today and the author has reached the conclusion that the story, although interesting, is likely to be apocryphal.

8.5 First demonstration: 1946 legislation, speed limit

The first of many demonstrations took place on February 13th 1946. Over two hundred people - including some of the island's legislative body - turned out to watch Lewis Lindley of Burnaby Cycles drive a New Hudson autocycle up and down Reid Street, Hamilton, the capital of Bermuda.

Fig 8.4 If a prototype Terrier had been built with pedals and folding foot rests, it would have looked something like this. 71 examples of this maroon painted 125cc model were imported in 1952. (Courtesy the Royal Enfield Owners Club)

On the 3rd April 1946 a temporary Act was passed allowing the general public to own and operate auto-bicycles with engines giving up to 2hp. Then, on the 3rd August 1946, the new Motor Car Act was passed. This Act included 'auto-bicycles', now giving certain cylinder bore maximum dimensions for various horse power classes, but still specifically excluding motor cycles. This Act also imposed many other regulations that are still in force today. One section of the Act required a governor to be fitted to the engines of all vehicles to prevent them exceeding the existing 22mph speed limit. This provision, however, was never enforced as no such device could be found. When the 22mph limit was set is not known, but it's believed that the figure may have borne some relationship to the speed of a trotting horse!

8.6 1949 legislation: pedals, engine power

On 22nd December 1949 the Motor Car Act was amended to reflect two classes of auto-bicycle:
(a) an 'auxiliary-bicycle' with a motor not exceeding 1hp.
(b) an 'auto-bicycle' with a motor not exceeding 2.1hp.
Both classes of machine still required the fitting of pedals capable of propelling the vehicle!

8.7 1951 and 1953 legislation: no pedals, weight limits

July 23rd 1951 saw the introduction of another Motor Car Act which replaced the much modified 1946 and 1949 Acts, but there were no significant changes regarding auto-bicycles. However, the definition of an 'auto-bicycle' was changed on March 3rd 1953, removing the need for propulsion by means of pedals on these more powerful machines, after a series of unfortunate accidents on the tight bends of the island's twisting and poorly surfaced roads. It was realised that the machine was far too unwieldy and dangerous in this form. The pedal requirement was, however, left intact for machines of under 50cc.

A 220lb (100Kg.) upper limit on the tare weight was also introduced at around the same time. Some sources have indicated that there may also have been a minimum weight limit in force, too, (presumably to further restrain performance). Lead tape is said to have been laid inside the wheel rims and an extra thick and heavy seat pan employed to get the weight up to the required figure. However, no evidence has been found in Bermuda or elsewhere to support this contention and it is considered unlikely.

8.8 1956 and 1973 legislation: capacity limited to 150cc

One further piece of legislation came with the Motor Car Act of 3rd December 1956, when it was decreed that 'auxiliary-bicycles' were to be of no more than 50cc

engine size (replacing the former 1hp limit) and the engines of 'auto-bicycles' were to be no more than 150cc in size (replacing the former 2.1hp limit).

Finally, on 25th May 1973 came the Act limiting engine size to 100cc of all newly registered machines. Fortunately, this legislation was not retrospective and there was a waiver for older machines with a capacity of up to 150cc. As a result Cubs, Terriers and many other machines these days continue to be the pride and joy of their owners, to the pleasure of all and the benefit of posterity.

8.9 Noise limit

Another interesting legal requirement was a noise limit, set at 93 decibels and tested locally on the island by government officials. In 1954 the Bermuda Transport Control Board had carried out a series of tests using a motor cycle with its rear wheel mounted and running on a set of rollers, simulating road use, while a noise meter took readings. The readings obtained, whilst perhaps satisfactory for the comparison of one machine to another, cannot have been truly objective as the tests took place inside a corrugated iron shed! However, it was found that a typical two-stroke produced 87 decibels whereas the Terrier gave a figure of 84 decibels. Due to the nature of the decibel scale, this meant that the two-stroke machine was some 30% louder than the four-stroke.

Provision was made by the Board for future legislation but it is not known when the 93 decibels limit was finally enacted. In the meanwhile, similar tests were being carried out at Meriden to ensure that the design of the machines would not prevent them from passing the local regulations once they had arrived on the island.

8.10 Legislation - summary

Past regulations put the limitations applicable to Terriers and Tiger Cubs into place. A 150cc engine (nowadays 100cc), 93 decibels maximum noise level, 220lb (100Kg), maximum weight and a speed limit of 22mph (35kph). The wearing of helmets has also been compulsory since January 1977. Only one car per household is permitted, (with a few exceptions such as doctors), but there is fortunately no limit on the number of motor cycles that can be owned and used by an individual.

One final point is that, at some time just after the second world war, it was decided to limit the number of colours allowed in machine livery: with the exception of black, only two other colours were permitted. Bermuda is a very beautiful and colourful island so, in view of today's garish motor cycle colour schemes, this very conservative decision was perhaps imbued with a greater wisdom than was realised at the time!

8.11 The first imports by Charles Young

Charles E.Young & Son imported their first motorcycle in 1946 and an early example of this new form of transport was paraded around the island and put on show for some six weeks. Orders were taken even before the legislation allowing importation had been finally passed. The first machines were 98cc Villiers engined Excelsior and New Hudson auto cycles, which sat on the dockside until the necessary legislation was passed by the island's parliament in early 1947.

This company became the Triumph main agent on the island but, from around 1963, things became difficult as labour problems at Meriden were affecting deliveries of motor cycles and spares. The supply of Terrier engine components - particularly 150cc barrels - began to dry up, making it difficult to maintain the machines. On top of that an order for twenty machines placed in 1964 was never fulfilled and Young & Son finally stopped importing Triumphs, switching allegiance to Honda with its new electric start 100cc and 125cc models. The connection between Meriden and the island was finally severed in that same year.

8.12 Engine configuration - '150' and engine types

All Terriers and Cubs were delivered with 150cc engines, even the T20SS models, with the exception of the last ten machines which came with 200cc engines and had then to be converted to the legal size by Youngs. The engines for the Tiger Cub models, although of 150cc, were stamped with 'T20' numbers and crankcases were stamped, '150' on the left side of the cylinder barrel plinth. Terrier engines were stamped with 'T15' numbers and did not have the '150' marking on the barrel plinth.

Engine configuration followed the general Terrier and Tiger Cub 'conventions':
1. A one-piece crankcase was used up to engine no. 57616; after that it was the two-piece type.
2. Up to engine no. 84268 there was a plain timing side main bearing. From no. 84269 it became a ball journal bearing.
3. Prior to engine no. 88347 a 'mushroom' type contact breaker was used; after that engines were the side-points type.

8.13 Engine configuration - barrel, pushrods and flywheel assembly

Terriers were already 150cc and required no modification. For Cubs the 150cc capacity was achieved by using some Terrier parts and a special 57mm bore cylinder barrel with an enlarged spigot, allowing it to fit into the standard Cub crankcase mouth. They also used Terrier pushrods, pushrod tubes and, up to engine no. 84268, the Terrier flywheel assembly.

From engine no. 84269 there was the problem of how to obtain the 58.5mm stroke necessary to arrive at 150cc capacity. Terrier flywheel assembly could not be used in a ball journal timing side main bearing crankcase. The solution was to use two Terrier drive side flywheels.

Fig. 8.13 A Terrier barrel for Bermuda modified with the enlarged spigot (left), and a normal barrel (right). (Courtesy James Hallett, Bermuda)

The left side was fitted with the long duplex chain mainshaft and the other side used the new short timing side mainshaft.

8.14 Cycle parts

Most Bermuda machines followed UK specification and colours, except for the changes required by the island's legislation. However, there were also some machines delivered between 1958 and 1960 that were painted in Aztec Red, the USA colour scheme. Some of the 1955 Terriers sent there were in the USA livery for that year, which was black with gold pin-striping on the wheel rims.

Recent news is that one Terrier from a 1955 batch of eight was actually a Tiger Cub. Still 'all original', it has T20 cycle parts and paintwork and a black painted 'Terrier' gear indicator medallion. Were the other seven machines the same, and other batches, too, perhaps? It certainly puts the figures in Appendix 6 into perspective!

8.15. Sales - general

A few machines went to the island's Health Department and also the police, but the majority was sold privately. Many went to Navy and Air Force personnel stationed at the American base in the north-east of the island. Due to favourable rates of exchange, US servicemen found it considerably cheaper to buy their machines and then ship them, probably free of charge, on board service transport back to the States. A number of Bermudan machines must have finished up in the USA, returning home with the owner at the end of his tour of duty, and then being modified back to 200cc once free of the island's regulations.

8.16 Bermuda today

Although Triumph lightweights have not been sold on the island since 1964, there is still today a strong and active group of riders of not only these machines, but also the many other small capacity makes and models imported into Bermuda over the years. Around eighty Cubs and Terriers are known to survive on the island and these machines are loved, regularly used, and clearly the pride and joy of their owners, being always impeccably presented and maintained. Quite right, too!

Fig. 8.14 1958 Aztec Red machines for the Bermuda Health Department. (Courtesy James Hallett, Bermuda)

Fig. 8.16 Some of the Cubs now surviving in Bermuda. Almost all machines in Bermuda have been painted in non-standard colours. They are regularly used, loved and cherished by their owners. (Courtesy James Hallett, Bermuda)

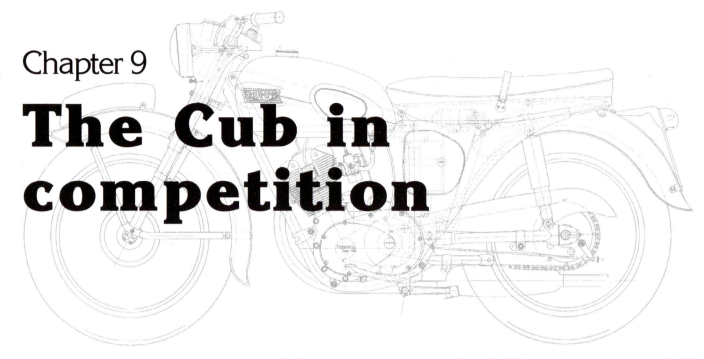

Chapter 9
The Cub in competition

9.1 General

In the UK the Cub was, among other things, road raced, sprinted, trialed, scrambled, hill climbed and grass tracked, whilst in the USA it was also raced indoors on artificially surfaced short tracks, and across country in desert enduro events. Cubs were also used in ice racing, the 'Wall of Death' and a Terrier was used by three Austrian acrobats

Fig. 9.1a A ladies 'Wall of Death' team on their 1954 Tiger Cubs. (The Mike Estall Collection)

Fig. 9.1b An ice-racing team with Tiger Cubs. The venue is the South Gate Arena, Los Angeles, circa 1955. (Courtesy R.E. Rogers, Little Rock, Arkansas, USA)

Fig. 9.2 'Jock' Hitchcock cleaning the oil tank cap on T15 107 during the 1953 ISDT in Czechoslovakia. Note that the centre stand has already suffered a partial collapse! (Courtesy EMAP)

in their circus high wire act. The author has also seen photos of a three-wheeled conversion! In events where outright performance was the prime objective, the Meriden lightweights tended to be fast but also somewhat fragile and their short wheelbase sometimes meant handling was a little too frisky.

In the UK the Cub's real forté was in trials events, particularly those where the sections required agility rather than raw mud-churning power. The combination of light weight, a short wheelbase and reasonable 'plonk',

Fig. 9.3a T15 107 in full trials guise after the treatment by 'Jimmy' Alves. Note the manual advance and retard lever on the left handlebar and its cable to the contact breaker. (Courtesy EMAP)

Fig. 9.4a May 1955, a racing Terrier, venue unknown. Note the twin leading shoe front brake and the home-made tachometer drive. (The Mike Estall Collection)

9.2 1953

combined with adequate top end power, made for a very good trials machine in all but the most power-draining conditions. In trials it was a watershed model, seeing the demise of the big four-stroke single and setting a trend towards the lightweight four-stroke machine, only pushed out of prominence by later foreign two-strokes.

A very early competition use of the new Meriden lightweight took place in October 1953. Jock Hitchcock, the well known Triumph dealer from Kent, took Terrier T15 107 (UK road registration RKO 88) into the International Six Day Trial as a private entry. Surviving until the fourth day, he retired with ignition problems. Later, the same machine became the factory's first attempt at a trials conversion on a lightweight when one of its top riders, P.H. 'Jimmy' Alves, got out the hacksaw and turned it into a trials machine on which he later had several class wins. The rear frame of the machine was widened and the mudguard mounting brackets moved to

Fig. 9.3c Jimmy Alves on his plunger Cub in the Scottish Six Day Trial. (Courtesy EMAP)

Fig. 9.4b The same 1955 racing Terrier from the drive side. (The Mike Estall Collection)

allow the use of a 4 inch tyre. The rear suspension was locked solid by the use of alloy blocks in place of springs, effectively making it a 'rigid' mount. Hitchcock also made a larger rear wheel sprocket, greatly reducing the overall gearing. The rather flimsy front forks were given some rigidity by the use of a thick Dural trials number plate bolted firmly across the top and bottom yokes.

9.3 1954

Alves gave the trials establishment quite a shock by putting up the best solo performance on T15 107 in the Colmore Cup Trial of February 1954. The engine was then bored out to 175cc and Alves rode it to a 151-200cc class award in the prestigious Scottish Six Day Trial of May 1954. He also later successfully converted a Tiger Cub for trials. This had more bottom end power than the Terrier and anticipated the later very successful Cub trials variants.

Even as early as 1954 a Terrier tuning kit was available in the USA. This kit, offered by Foreman Engineering of Borger, Texas, and costing $41.65, contained a 9:1 high compression piston, stronger valve springs and an oversize inlet valve with a stem waisted behind the head. It was recommended that the cylinder head should be ported and the tappets reground to a $1^3/_4$in radius. A new camshaft made by Harman & Collins Co. in California, later to be known as the 'Big Bear' cam, could also be fitted. Power went up from 8bhp at just over 6000rpm to 12.5bhp at 7000rpm, or 14.5bhp

Fig. 9.4d Doug Beasley's twin ohc Terrier fully clad in its fairing. (Courtesy EMAP)

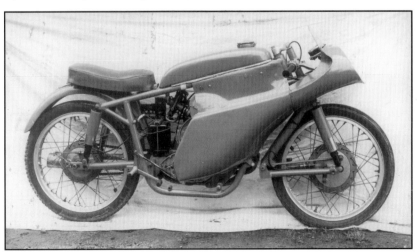

Fig. 9.4c Doug Beasley's 'double knocker' Terrier with its MV frame and Earles type forks. (Courtesy EMAP)

Fig. 9.5a Ken Heanes' 1956 ISDT gold medal winning machine. Note the extra finning on the inlet rocker box, brought into general use at machine no. 26276. This machine is worthy of close attention for the many other useful features shown. (The Mike Estall Collection)

at 7250rpm with special fuel. Top speed attainable using this kit was near to 90mph.

The special fuel, for a five gallon mix, consisted of three gallons of 91 octane aviation petrol, one gallon of benzole, 3/4 gallon of methyl alcohol, 40cc of acetone and 30cc of degummed castor oil as an upper cylinder lubricant. It was also recommended that ignition timing was advanced by 6° and the main jet size increased to 130 or 140.

9.4 1955

Available for the Cub in the USA in 1955 was a 10:1cr Robbins piston, made in Los Angeles. It had twin 'V' strengthening struts from the gudgeon pin boss to the crown, incorporated a 1/8in wide oil control ring and came in sizes from standard to +0.100in.

The factory may have made ten 125cc cylinder barrels with oversize square-shaped fins, for use with Italian Mondial pistons in machines built for racing or for trials. The provenance of this statement is open to question and the year not known, but is believed to have been in the mid-fifties. It is also considered doubtful because E.T. was usually against time being wasted on projects such as this.

A double overhead cam racer with a Terrier-based engine was put together by the well known Coventry specials builder Doug Beasley. The crankshaft had steel flywheels and a roller timing side main bearing with ball and roller bearings on the drive side. The conrod was carved by hand and Beasley cast his own cylinder head and camboxes. The bore and stroke was 53 x 56mm, giving 123cc, and the compression ratio was 9:1. The engine was expected to go to 11,500rpm and it was put into a steel-tubed frame coming from a 125cc MV Agusta fitted with Earles forks. History does not show how successful, or otherwise, this venture was ...

Swinging fork conversions and Earles type front forks were available from Bill Martin of Sacramento, California.

Fig. 9.5b Heanes' ISDT machine from the drive side. The machine was T20 23411, sold in December 1962 to J.A. Hitchcock in Kent. (The Mike Estall Collection)

Fig. 9.5c A 175cc conversion kit - Terrier barrel with adapter to fit it into a Cub crankcase and 3TA piston. (The Mike Estall Collection)

Fig. 9.5d A.J. Dolphin's Featherbed-framed trials Cub. A lot of frame for a little engine! (Courtesy EMAP)

Fig. 9.6a A factory racing conversion with total loss electrical system, remote float 276 carburettor, exhaust pipe clamp and tachometer drive. (The Mike Estall Collection)

9.5 1956

A heavy duty clutch kit, racing valve springs, an oversize inlet valve and a range of gearbox and rear wheel sprockets were available for the amateur tuner in the USA. As well as the Harman & Collins 'Big Bear' cam, there were by 1956 about half a dozen other cam grinds from this company, each with the valve timing figures stamped on its end face.

An American Bates flywheel magneto could be

Fig. 9.6b Johnnie Giles' works scrambler, June 1957. Note the head-steady, a guide for the rear chain and adjustable suspension units. The fuel tap is in an unusual position too! (The Mike Estall Collection)

Fig. 9.7 A works prepared trials Cub from January 1958. With 20mm Zenith carburettor and a head-steady. The gearbox cover has been split for easy access to the clutch cable end. (The Mike Estall Collection)

retro-fitted in place of the alternator and HT coil. Bates recommended that the auto advance unit be removed and the engine run on fixed ignition timing. By 1959 three different types of this magneto had evolved; one for single row primary chain engines, one for double row and another for use on engines after no. 48340, where the rotor keyway location had been changed. The enterprising Americans also found that flywheel magnetos from chain saws, outboard motors, etc., could be substituted for the Energy Transfer system.

American racers often improved the oil feed to the rockers by taking it directly from the oil pressure test hole at the front of the crankcase.

Cubs were ridden by George Fisher and Ken Heanes in the October 1956 ISDT in Bavaria, where Heanes won a gold medal in the 175cc class, the T20 engines having been converted by the factory to that capacity. Heanes' machine, with British registration UAC 377, was given the serial number TR20 23411 and sold in 1957 (long

before the first 'proper' TR20 in 1962).

The conversion to 175cc was done by boring a Bermuda Cub barrel, with the enlarged spigot so that it fitted in the Cub crankcase, to 58.9mm and using a 3TA piston and rings. Later 175cc conversions were achieved either by using a Terrier barrel with a spigotted adapter in the crankcase mouth and a 3TA piston, or an eccentric crankpin for other capacity sizes. The two ISDT machines also used swinging fork rear suspension for the first time on a Cub, and many other features were used on the T20C introduced a few months later.

These ISDT machines were full of good ideas. All control cables were duplicated and strapped into place alongside the working cable. The timing side outer cover was cut allowing the rear part to be removed, giving easy access to the clutch cable end. The inlet tract was lengthened, as was the centre stand. There was a head-steady bar to the frame top tube and adjustable rear suspension units from a twin cylinder machine. The oil tank sprouted a large froth tower at the front and there was a D-shaped air cleaner sited behind the battery box. A CO2 bottle, strapped to the rear subframe with its tyre adaptor neatly clipped to an extended rear number plate bolt, completed the ensemble.

Harry Woolridge built a Tiger Cub trials bike with an additional 'outside' flywheel in place of the alternator rotor, and a total loss battery powered ignition system. It apparently would 'plonk' its way up anything but consumed drive side mainshafts at an alarming rate.

One competitor in the Scottish Six Day Trial, A.J. Dolphin, put a Tiger Cub engine in a Norton Featherbed frame with leading link front forks!

9.6 1957

Both riding Tiger Cubs, Ken Heanes won a first-class award in the 'Scottish' and Artie Ratcliffe won the 'Scott' trial.

Arthur Wheeler of Epsom, Surrey, built a Cub with Armstrong leading link forks and a magneto sited over the distributor position. A similar set-up was used by at least one French rider in grass track events who fitted a WiCo competition magneto which was driven by a chain from the end of the camshaft. Using a magneto in the distributor position may have been electrically successful, but the torque of the magneto could very quickly wreck the skew gears if they were used to drive it.

9.7 1958

Tiger Cub riders were 3rd (Ratcliffe), 6th (Roy Peplow) and 10th in the Scottish Six Day Trial.

Meriden-made performance parts became available, including a high compression piston, interference valve springs with alloy caps and new cups, the 'R' cam, large inlet valve and an exhaust tailpipe extension. Interference of the inner valve spring with the outer spring helped dampen resonance or spring surge, which could badly upset valve timing at high engine speeds if it was not properly controlled.

9.8 1959

Roy Peplow won the Scottish Six Day Trial on a Tiger Cub. It was the first time this premier event had been won by a machine of less than 250cc capacity. Ray Sayer was fourth and Triumph also ran off with the manufacturer's team prize, Artie Ratcliffe being the third rider. Triumph Cub riders also took the team prize in the West of England trial later the same year. Ray Sayer won the Allan Jefferies trial on a Cub in 1959, and again 1961 and 1964.

Bill Martin from Burbank, California, rode a petrol-fuelled streamliner on the Bonneville Salt Flats, Utah, on 26th August 1959, setting an AMA 200cc Class C one mile record with an average two-way run of 139.82mph, including a best one-way speed of nearly 150mph.

Fig. 9.9a An interesting little sprinter. Note the magneto and what looks like a Wal Phillips fuel injector. (Courtesy EMAP)

Fig. 9.9b Roy Peplow on his trials Cub in the winter of 1961/2. (The Mike Estall Collection)

US riders could buy an efficient bolt-on oil filter kit with replaceable cartridges. The filter unit was located under the seat, interrupting the oil tank return line, and was highly recommended by the Triumph Corporation, especially for machines used in dusty conditions.

Another racer built by Doug Beasley used the light alloy frame from Les Graham's 1953 125cc TT winning machine. It had an engine based on a Tiger Cub unit but with a Velocette MOV cylinder barrel and a Triumph TR6 Trophy piston giving a capacity of 249cc. This latter machine was geared for 7500rpm at 100mph, weighed 162lb and was ridden with some success by none other than Triumph tester supreme, Percy Tait.

The Cub tuning 'Bible' came into being this year. Meriden's *Tuning Information Bulletin no. 4*, affectionately known as 'TIB4', showed the average owner how to put together all the performance parts then available such as high compression pistons, sports camshaft, interference valve springs with their cups and caps, tachometer, close ratio gears and remote float carburettor, plus information on porting, overall gearing, exhaust systems, and so on. In other words, everything necessary for the ordinary boy-racer to take his machine towards self-destruction in a blaze of high performance glory!

The Bulletin appeared at a time when the one-piece crankcase was extant. Fortunately most of the precious information contained therein was still applicable when the much stronger two-piece cases arrived on the scene. As each of the later developments were applied to the Cub engine, so engine/gearbox unit durability increased and the tendency for mechanical suicide diminished. In the end, and if a late type engine was used, TIB4 became the means of producing a pretty good little performer.

9.9 Late-fifties and early sixties

By the late-fifties Triumph had taken on, or were about to take on, several 'works' riders whose names became synonymous with Tiger Cubs and award winning - Gordon Blakeway, Scott Ellis, Paul England, Gordon Farley, George Fisher, Johnnie Giles, Ken Heanes, Roy Peplow, Artie Ratcliffe, Malcolm Rathmell, Ray Sayer and Dave Thorpe. The glory days had begun modestly enough in 1954 and lasted for several years, reaching a zenith in 1959, but a few short years later, by the end of 1966, Triumph had retired from trials and it was all over.

American events such as the Catalina Grand Prix, Big Bear Run and the like, were dominated by Triumph Cubs and Terriers in the 200cc and 165cc classes up to the end of the fifties, but the Meriden machines were then overtaken by Italian lightweights. In short track racing Cubs and Terriers with purpose-built rigid frames were very successful until the Harley/Aermacchi overhead cam singles and Bultaco two-strokes came along. It was a similar story in road racing but here it was the Ducati and Parilla machines that put an end to the Tiger Cub's dominance.

9.10 1960

At the end of 1959 season, a 'Woods' kit for T20S models was a bargain offered to off-road riders by the Triumph Corporation of America. The kit included a $^7/_8$in bore Amal 332/7 carburettor, air cleaner, low compression piston, standard camshaft and a 58 tooth rear wheel sprocket. The retail price of the parts was nearly $55 but the price to dealers was under $30. If bought with an 'end of season' T20S, the kit was further reduced to $22.40 and the bike itself reduced by another $35.

Also available in the USA at around this time was a roller ended cam follower which was extensively used by racers and speedway riders.

A 242cc grass track machine with a Cub-based engine was built by Fred Wells of Ilford, Essex. This time the bore had been opened to 65mm and the stroke increased to 72.5 mm by means of an eccentric crankpin, the engine having a 13:1 compression ratio. Methanol fuel was fed to the engine via a $1^1/_{16}$in TT Amal through a large inlet valve cylinder head. The machine featured a much modified Royal Enfield Ensign frame with the top tube converted for use as an oil tank and undamped swinging fork rear springing.

9.11 1961

Big end failures in USA competition circles were thought to be caused by excessive oil pressure. This led to more than one after-market manufacturer providing adjustable pressure relief systems.

Available in England were various after-market tuning accessories, including a 65mm bore alloy barrel for £9.15.0, (£9.75), and 8.5 or 10:1 pistons at £10 from B & C Pearson, near Birmingham. This brought capacity up to 212.37cc.

All-alloy tubular pushrods for use in racing were available in the USA at $2.70 each. The standard steel cupped alloy pushrods were alright up to about 8000rpm, but beyond that the tubular type was recommended by Triumph Corporation as being both stiffer and lighter.

Fig. 9.10 Mr and Mrs Ken Maely, long-time steel racing shoe maker, and Jim Skinner's Junior speedway bike, 'Saturday Nite Slider'. (Courtesy Dan Andrade, Washington State, USA)

Fig. 9.12a Rayer Sayer ready for action at the 1963 Scottish. See the early use of the pivoting kickstart lever. (The Mike Estall Collection)

Fig. 9.12b Sayer in action in the 1963 SSDT. (The Mike Estall Collection)

Fig. 9.12c The 1964 Metropolitan Police trials team on their TR20 machines. The rider's names are not known. (The Mike Estall Collection)

At least two after-market suppliers made 250cc conversion kits in the USA which were recommended by TriCor. Hertting's Triumph Sales from Milwaukee, Wisconsin and Dwain Taylor of T & M Motors in Albany, Georgia could both supply kits in 1962. These kits could be fitted to 1960 or later Cubs and converted the bore to 71mm.

By 1963 one or other of these kits was being marketed by TriCor itself with a modified 250cc cylinder head and a pushrod tunnel with external oil drain tubes. A 71mm 11:1 or 12.5:1 compression piston with +0.010in and +0.020in oversizes was used, and the connecting rod was modified to take its larger diameter gudgeon pin. There was also a modified flywheel assembly allowing the use of the latest $1^5/_{16}$in diameter big end with the bush type timing side main bearing. The whole kit sold for about $137, or slightly less if certain items were exchanged. This conversion was legal in AMA Class 'A' or 'C' short track racing or USMC 250cc events but it was not legal for TT or road racing which required machines to have a bore and stroke as catalogued by the manufacturer.

Dalio's Motorcycles of Fort Worth, Texas, produced what was known as the 'Texas barrel'. Using a 69mm 5TA piston, it had the fins at the front cut away to allow room for the frame down tube and at the rear making space for the contact breaker. The pushrod tube was totally enclosed by the barrel finning. The spigot into the crankcase mouth was a little shorter than standard to allow room for a longer stroke flywheel assembly.

The larger bore size of the Texas barrel cut through the normal rocker box oil drains, so, instead of using external piping as in some other big bore kits, oil drained down through modified drillways in the cylinder head. From there it went into the cylinder barrel stud cavities, and down into the crankcase through holes drilled into the bottom of the cylinder head studs.

Finned, cast aluminium rocker covers were available in the USA. From August 1962 these become the standard factory fitting on all machines. Special 'Grant' piston rings became available there, too, for use with high compression pistons.

9.12 1962/3

A special single valve spring was now on sale in the USA for use with new top and bottom collars and alloy collets. An inner valve spring was not to be used with this special spring and the kit cost $5.75.

Also available in the USA was an eccentric rocker ball pin. This reduced the distance from the rocker shaft to the centre of the ball pin from the standard 27mm to 25mm, increasing the rocker ratio from 1·1:1 to 1·2:1, thus gaining a little more valve lift from the cam.

9.13 1964/5

Comerfords of Thames Ditton, Surrey, sold a kit of alloy parts for Cubs. It included an alloy top yoke for heavyweight forks, front and rear brakeplates, oil tank and air filter, at a total cost of £17.2.6d (£17.13). The alloy parts were made by trials riders Murray Brush and Mick Bridger, who rode Cubs for the trials team of Kent dealer Jock Hitchcock. Also available was a 'Lyta' BSA Victor style alloy petrol tank made by Colin Hitchisson.

9.14 1966 onwards

Another accessory was the 'Cycle Sno-Go' made in Sacramento, California. This contraption could be fitted to many lightweight machines, including Tiger Cubs, enabling them to traverse snowfields or sand at up to 20mph. It employed a pair of skis attached to the front wheel and twin caterpillar-type rubber tracks fitted in place of the rear wheel. The contact area was 990 square inches and it gave a per-square-inch weight of only six ounces. The device weighed 130lb, took thirty or forty

Fig. 9.12d A 'short track' 250cc engine with a 'Texas' barrel. Note the various rocker covers on the table and three pistons - a normal 7:1, a high compression 9:1 and a racing 10.5:1. Also shown are some after-market clutch plates. (Courtesy James Hallett, Bermuda)

minutes to fit and cost $304.16, or slightly less than half the cost of a complete Mountain Cub. Marketed by Johnson Motors, it is not known how many (or even if any!) were sold.

Bored to 67mm, 226cc machines were ridden by Gordon Farley and Paul England in the Scottish Six Day Trial. Roy Peplow rode a 199cc machine and Malcolm Rathmell a 148cc engined machine in the 150cc class.

Companies like Jacksons, Cheetah, Cheney and Sprite made frames into which owners put their own Cub (or Villiers) engine, also attaching the appropriate cycle parts.

9.15 The 'Comerfords Cub'

In October and November 1967, twenty Mountain Cubs were delivered to Comerfords, the well known Triumph main agent in Surrey. Another forty machines were delivered early in 1968 and the sixty machines, converted to trials specification, became forever known as 'Comerfords Cubs'. It has been suggested that the machines came from a cancelled US Army order, but no evidence has been found to support this theory.

Comerfords had engaged Gordon Farley, the ex-Triumph 'works' trials rider and superb spanner-man, to work under its Competitions Manager, Reg May, using his specialist knowledge to convert these machines into copies of his 'works' trials bike. Previously, Farley had ridden a Cub very successfully for Kent dealer Jock Hitchcock, his machine utilizing the many alloy parts made by Murray Brush, Mick Bridger and Colin Hitchisson then being marketed through Comerfords. When he later went to Triumph, Farley became subject to their policy of only using machines that a customer could buy from the factory. So, from the machine specification point of view, he took a step backwards when he accepted a 'works' ride.

However, when Triumph made the last batch of TR20 machines in February 1965, knowing there would be no more, they allowed Farley all the alloy parts previously denied him, thus bringing his machine up to a more competitive specification. It was this machine that was used as the pattern for the Comerfords Cubs.

By the time all this happened, Farley had left Triumph to ride for Greeves, being taken on by them on 1st November 1967, and it was from them that he went to the Thames Ditton dealer's competition department.

The converted machines were advertised as being exact replicas of Farley's own factory model that he had used up to the end of Triumph's interest in trials. Although Triumph had announced its withdrawal from trials competition and disbanded the team on 18th October 1967, Farley must have been left with a bike and so the conversions were done with Triumph's full knowledge and consent.

The machine was a Mountain Cub with modifications: there was a 21in front wheel, a 'Lyta' BSA Victor-style alloy petrol tank bolted to an additional frame strengthening cross-brace from top of the seat tube to the headstock, and an alloy oil tank. The Mountain Cub wide ratio gearbox was used with a 17 tooth gearbox sprocket, but a 64 tooth sprocket on the rear wheel replaced the

Fig. 9.14 The 'Sno-Go'. This looks like a heap of fun. The rider is Dale Martin, son of Bill Martin (see Ch 7.17). (Courtesy R.E. Rogers, Little Rock, Arkansas, USA)

original 54 tooth item. This saved the additional expense of completely stripping the engine to fit a smaller gearbox sprocket. The carburettor was changed to a $^{25}/_{32}$in choke 375/44 Monobloc (the same as the TR20 model), instead of the original 376/314 of $^{15}/_{16}$in bore.

There was a small trials seat, steel Sammy Miller type handlebars with a four inch lift (later 'Renthall' type), and alloy mudguards front and rear. Flat section footrests were brazed in place a couple of inches below the swinging fork pivot, the right hand side one folding

Fig. 9.15a Gordon Farley with his 'checkerboard' hat at the Scottish, May 1967, on his 'works' Cub. This is the machine used as the pattern for the Comerfords Cubs. (Courtesy EMAP)

Fig. 9.15b Gordon Farley again on his 'works' Cub. (Courtesy EMAP)

to allow the standard cottered kickstart lever to swing freely. Farley's original mount had used the splined type of kickstart lever, but the cottered type was retained by Comerfords on the grounds of cost.

The speedometer was relocated in the crook of the upswept exhaust pipe, attached by a bracket to the front engine mount, and all lighting and headlamp brackets were removed. In addition, an alloy silencer and air cleaner were used. The alloy top yoke, oil tank and front and rear brakeplates then available from Comerfords could be fitted on request to save weight.

Although the machine was advertised as being an exact replica of Gordon Farley's 'works' machine, this was not strictly correct. The Comerfords Cub used a standard 17 tooth gearbox sprocket whereas the factory bikes would have used smaller sprockets down to 13 teeth. It also retained the Mountain Cub steel chainguard, brakeplates, cast iron top fork yoke and standard kickstart lever.

The complete converted machine sold in November 1967 for £225 but was reduced in January 1968 to £210 (the Mountain Cub was just under £200 at the time). These days Comerfords Cubs are widely appreciated and in demand. The Tiger Cub and Terrier Register has the recorded details of eleven machines currently in existence from the original sixty.

Fig. 9.15c The Comerfords Cub. One of sixty machines modified by Gordon Farley at Comerford's Thames Ditton premises. (Courtesy EMAP)

Chapter 10
Military, Police and the Utilities

In October 1959 Terrot was the largest motor cycle manufacturer in France, and also the French main agent handling all sales, distribution and servicing of Triumph machines. An announcement was made in the motor cycling press that all machines supplied to France were to carry a Terrot badge in addition to the Triumph motif. In 1960 Terrot was taken over by Peugeot, which, up until then, had been France's second largest motor cycle manufacturer, but Terrot retained its main agent role up to the end of 1962.

In January 1963 CGCIM (Comptoirs Généraux du Cycle et de L'Industrie Méchanique, 17, Rue du Débarcadère, Paris 17), which was owned and headed by Messr. Guy Peyron, became the French agent for Triumph machines, taking over this position from Terrot. Messr. Peyron was a son-in-law of the Peugeot family and

Fig. 10.2a Bastille Day celebrations in Saumur, Near Nantes, France, 14th July 1977. What is the appropriate collective noun? A 'Crackle' of Cubs? Perhaps a 'Clatter' of Cubs? (Courtesy Decker, Saumur, France)

107

Fig. 10.2b **More Bastille Day activities from Saumur. An appropriate moment for a caption competition perhaps? (Courtesy Decker, Saumur, France)**

although he now headed the Triumph main agency in Paris, he was a businessman, not a motor cyclist and knew little about motor cycles.

10.2 Orders and first deliveries

In May 1961 a T20T machine had been sent to Terrot, kept there for eighteen months and then returned to Meriden to be dismantled. It was probably the same as the military pattern T20T machines delivered to the British Army at around that time. The French military was seeking a replacement for its 175cc Peugeot machines so representatives would have had photographs and specification details of a military Cub to see when they made their enquiries at Peugeot or later on at CGCIM.

Another Triumph dealership 30km south of Paris was run by a Messr. Valdevit. He was asked by Guy Peyron to act as his advisor in any discussions with the French military or with Meriden. Messrs. Valdevit and Peyron would no doubt have also visited Meriden to discuss technical details, and order and delivery schedules for the machines.

So, in June or July 1964, the first machines were built for the French Army and sent to CGCIM. Not long after this Messr. Valdevit began selling Honda motorcycles, which pleased neither Messr. Peyron nor Triumph. Valdevit eventually became disconnected from Army dealings, although did remain a Triumph and BSA dealer until 1973. He did not reappear in connection with the French Army Cub until their disposal sales in 1975.

The first machines were designated T20WD and incorporated earlier experience gained not only from the 1961 British Army T20T machines, but also from various trials machines that had been prepared by the Meriden Competition Department. Orders for more machines followed during the next few years, including one of 570 Cubs reported in the 18th May 1967 issue of *Motor Cycle*, but the precise ordering pattern is not known. The total order has been reported as being for 7000 machines, but

delivery was overtaken by events and the order was never fulfilled.

10.3 Technical details

The French Army Cub was basically a side-points engine with T20T internals fitted into a Mountain Cub chassis. There were further additional features designed to protect it from the rigours likely to be meted out by the average French 'squaddie'. These included strengthened footrests, heavier gauge control cables and a $^5/_{16}$in wide rear chain, gearbox and rear wheel sprockets instead of the usual $^3/_{16}$in size. The machine was fitted with a crankcase undershield and full road lighting. Once the machines arrived in France they were equipped with French made black leather panniers on a tubular frame bolted over the rear wheel, a crashbar and a left side handlebar mirror.

The standard cam was used, the low compression piston, a 'square' cylinder barrel and head (although the first 82 machines delivered in 1964 would have utilized the 'oval' head in use at that time), and sports valve springs. A slider-block driven oil pump was fitted on all but the 1964 machines, a $^{25}/_{32}$in choke 375/44 Monobloc carburettor (same as the TR20 model) and wide ratio gears. Energy Transfer ignition was used on virtually all of them and they were mostly fitted with the caged needle roller big end assembly. However, the 1964 machines used the large diameter plain bush type big end, and a few of these early machines also had AC/DC rectified coil ignition systems with battery.

The oil pipe exit tube from the oil tank was moved rearwards to prevent the oil pipes from rubbing on the crankcase and chafing. The front frame down tube was double gusseted at the root of the swan neck, and the rear sub-frame had a bracket for a high-level silencer, although the low-level system was used with a silencer and mute. The tube which carried this bracket was straight, top and bottom.

10.4 Colour scheme

The machines were painted all over in NATO anti-infra-red matt Green and the dualseat was black all over. The tyres were sometimes painted with white sidewalls for ceremonial parades, Michelin being favoured for on-road work and Dunlop Trials Universal for most off-road activities. The wheel rims, exhaust system, handlebars and levers were chrome-plated but otherwise it was drab green everywhere. Nevertheless, it was quite a handsome machine when in pristine condition.

10.5 Machine identification

The frames bore the usual 'F4421' headlug casting number stamped on the left hand side of the steering head but it was often stamped upside-down. The frame number was sometimes stamped over the top of the casting number.

The machines were variously described in the factory despatch books as models T20WD, T20M.WD, T20SM or T20M. The frame and engine numbers actually stamped on the machines usually began with 'T20SM' or 'T20M', followed by the serial number and sometimes also by a letter 'F', showing it was a French Army engine, or '17T', indicating the gearbox sprocket size.

There are some machines now in existence with engine numbers having six digits such as no. 100031 or 100240. These numbers do not appear in the surviving factory records but they are recorded in the Tiger Cub and Terrier Register. It is possible that these were engines built in France using new crankcases as it was not BSA Group policy to use six digit serial numbers.

10.6 Machine maintenance

Only the simplest basic maintenance or removal and refitting of complete engine units was permitted in French Army operational units. All maintenance was given an 'Echelon' (Stage) coding by the Army with numbers 1 to 5. Echelon 1 was checking tyre pressure or performing an oil change. Echelon 2 involved the checking and adjustment of carburettor or ignition settings, spark plug and tyre changes.

Anything beyond that had to be done by Army workshops at ERM (Etablissement de Réception du Materiel), Muret, near Toulouse, one of the many places where all Army vehicles were prepared prior to delivery to the units and the place where all the Cubs were sent for service and repair. Here, Echelons 3, 4 and 5 were carried out and a metal plate with the appropriate coding attached to the crankcase mouth.

Echelon 3 meant that the engine had been opened; *i.e.* the head or a side cover had been removed. Echelon 4 indicated that the engine had been rebuilt to manufacturer's standards and Echelon 5 showed that the whole machine had been subjected to a complete rebuild. Renovated engines would be stored in a wooden box for later fitting to other machines coming in for maintenance. Alternatively an engine might be returned in its wooden box to another unit, which may not necessarily be the one from which it came. This is why French Army Cubs nowadays almost never have matching frame and engine numbers!

The Muret workshops started operations in November 1965 but their first work on Cubs did not take place until 1969. How maintenance was done before that is not known. Their records show that they repaired 28 Cub engines in 1969, 15 in 1970, 244 in 1971, 310 in 1972, 225 in 1973, 70 in 1974 and 185 in 1975. In 1975 they also received 50 engines which were eventually deemed beyond economic repair and were probably scrapped.

The number of engines shown as having been repaired or scrapped - 1127 - represents a very high proportion of the total number of machines delivered, which is currently estimated at 1480. This may give some idea of the treatment the machines received from the French Army riders and also the inability of the Cub to withstand such assaults upon its person.

10.7 Problems with frame breakages

During their period in the French army the Cubs were regularly serviced but inevitably often had a very rough time. Frame breakage under the headstock was not

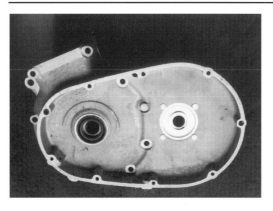

Fig. 10.9 The clutch seal carrier plate as fitted to some spare French Army crankcases. (Courtesy Jean Caillou, Paris, France)

unknown, particularly, for some reason, on machines fitted with trials tyres when ridden on tarmac. Frames cracking at the front engine mounting were also reported, possibly the result of poor brazing, and many machines had to be sent to Muret for front frame repairs.

10.8 Problems with electrics

It has been noted that, possibly due to cost constraints at Lucas, there was barely enough wire in the alternator windings to give sufficient current at kick-over speeds. This gave insufficient kilovolts at the sparking plug and sometimes made starting less than easy. A few more metres of wire in the coils might have made all the difference. It had also been found that with the ignition timing set at the recommended 10° of advance, the spark was required before the coil had produced peak current.

10.9 Problems with oil leaks

The Cub in general (and the French Army model was no exception) was always inclined to exhibit a degree of incontinence, depositing oil that had leaked through the seal behind the clutch. The problem was that as the bronze high gear bush wore, allowing the gearbox mainshaft to move around, the soft rubber of the seal would wear away the harder material of the bush until the seal could no longer do its job properly. To deal with the problem ERM Muret discarded all the standard E3932 labyrinth seals and used a locally made alternative, but there was another more serious problem that negated any effort to improve the seal.

A fault at the gearbox mainshaft centre, probably arising from either the poor state of the crankcase dies or the boring jig for the gearbox and engine bearing housings, affected large number of crankcases made for the French Army. Here, the gearbox high gear bearing housing in the timing side crankcase casting was not concentric with the clutch seal housing in the drive side. After machining the two housings could be 'out' by as much as 0.025in (0.635mm). The result was rapid and excessive wear on the clutch oil seal and a major oil leak at this point.

Stan Truslove and another Cub specialist from Meriden were sent to Paris to deal with this and other problems. On arriving home they made three experimental cases where the clutch oil seal was mounted on a separate plate secured in the correct position by three screws, in the same manner as on the Triumph twin cylinder 'C' range. This modification was accepted and all remaining crankcases were altered, but with the plate permanently fastened to the crankcase by four rivets. It would have been far more useful if the plate had not only been removable, as in the three experimental cases, but also large enough to enable the gearbox sprocket to be withdrawn through the aperture, allowing it to be changed without having to strip the whole engine.

10.10 Problems with centre stands

Some machines were sent to France with the ordinary sports Cub centre stand instead of a reinforced type specially developed for the French model. The centre stands were also found to be catching the silencer and so had to be returned to England and exchanged for the modified type.

A batch of these modified centre stands was made and sent to Paris, but frantic telephone calls from France soon revealed that the stands had not arrived. Somehow they had been inadvertently shipped to Canada and the main agent there must have wondered what on earth was going on when he received a crate of drab green stands of non-standard design! They had to be returned rapidly to England and then finally re-shipped to France.

10.11 Problems with overheating

The overall gearing on these machines (a 17 tooth gearbox sprocket and 58 teeth on the rear wheel) was fine for convoy duties, short road mileage and cross-country riding, but was too low for fast road work: if pushed too hard for too long it could result in overheating. ERM Muret reported many instances of piston seizure on machines which had been ridden in this fashion.

10.12 Cessation of contract

At some point Messr. Peyron discovered that the machines were being built at Small Heath, not at Meriden, as they had been when the original order was placed and as he had believed they would all be. All the negotiations had taken place at Meriden and the contract had been signed there too; as far as the French were concerned they were buying a Triumph machine made at the home of Triumph which was Meriden! They cared nothing for Harry Sturgeon's new domestic arrangements and Messr. Peyron became most unhappy when he learned that the French machines were being built by BSA at Small Heath.

Another serious cause for concern was that although the machines were being built at Small Heath, all the warranties were held by Triumph at Meriden. BSA may not have been too concerned about build quality (a subject covered in Chapter 6), and some machines were delivered to France in something less than a satisfactory state. Some were even found to have been delivered without an oil pump or with foreign bodies inside the crankcases - like the odd loose ball bearing and, on one occasion, a ballpoint pen!

After delivery of roughly the first thousand machines, and when the problems with frames, oil leaks and electrics had come to light, the poor build quality of the Small Heath Cubs had become clearly evident. Messr Peyron became very anti-BSA, developing a completely new attitude towards the Cub and it soon became his most fervent wish that the contract be terminated as quickly as possible. The French, therefore, decided to demand that the outstanding balance of the order be delivered at once; i.e. all six thousand remaining machines. They knew it would be impossible for BSA to comply, providing legitimate grounds for them to beat a retreat.

Fig. 10.17a A 1961 War Deptartment T20T machine in Deep Bronze Green. (The Mike Estall Collection)

10.13 The French Army Cub delivery total

The BSA despatch books show the following deliveries of likely military machines to CGCIM, Paris. Various model types as described as being sent - T20WD, T20SM,

Calendar year	Machines
1964	82
1965	-
1966	722
1967	303
1968	100
1969	144
1970	51
Total	1402

T20M, T20M.WD and T20B Super Cubs:

Added to these 1402 machines must be a further 117 whose serial numbers are not recorded in the BSA despatch books but which have appeared in the 'Domaines' (the full title of this organization is 'Service Central de Publicité des Domaines') sales records (see 10.15 below). Of these additional 117 machines, 5 are still in existence, recorded in the author's Tiger Cub and Terrier Register. The total number of French Army Cubs that can now be accounted for has therefore risen to 1519.

By comparison, the BSA quarterly production summaries show only 916 French machines built between August 1965 and July 1969. Added to this number must be the 82 built at Meriden in 1964, making a total of 998 machines. However, one page of these summaries, covering the three months from 22nd July to 20th October 1967, is missing. At the normal production rate of 100 to 150 Cubs per week, it is feasible that up to another 1500 machines or more might have been built in this three month period, so these records do not provide a true picture.

It has to be stated that the BSA despatch books contain many hundreds of unused machine serial numbers. There are gaps in the number sequences and no details are shown on these lines of entry. However, the Tiger Cub and Terrier Register now has on record some machines that fit into these unfilled spaces and it has become apparent that the factory records are less than complete. It is therefore now impossible to identify and account for every French Army machine and the only hope of ever completing this task may lie in France, where perhaps records there might provide further information.

As time goes on, more machines not shown in the factory records will be recorded in the Tiger Cub Register, but without the discovery of new archive material it's unlikely that the final number of French Army Cubs actually built and delivered will ever be accurately established.

10.14 The French Army Super Cub?

The BSA despatch records show one hundred T20B Super Cubs delivered to CGCIM, Paris, in July 1969. Although the French Army records show no T20B

machines in its various disposal sales, the author owns a brand new T20B frame which is painted in military green. No other military use of this model is known and the inescapable conclusion is that these one hundred machines were destined for the French Army. The colour of the frame, the large quantity of machines involved and the timing of delivery make it unlikely to have been an ordinary civilian order, leaving the French Army as the most likely recipient.

10.15 The French (dis)Connection

By 1975 the French had begun replacing the Cub with Honda CB250 twin cylinder machines and also Peugeot SX 8 models. The greater majority of Cubs had been used mainly for despatch riding, convoy duties and as a general runabout in mainland France, but a few had gone to Djibouti, in what used to be French Somaliland. It is not known how many went there, but some time in the seventies all the spares - and probably all the remaining machines there - were disposed of, consigned to a watery grave in the Red Sea.

Wherever they were Cubs were used until they broke and reached a stage of decrepitude that was beyond economic repair. At this point the Army ceased repairing them and they were gradually gathered together from all corners of the French military empire and auctioned off at the various 'Domaines' sales.

In March 1975 the first 40 Cubs were sold via 'Domaines'. (Even today, this is the service that buys and sells all materials used by French governmental administrations, including the armed forces.) There were many other sales over the next few years, sometimes with only one Cub but usually with many more, including one sale at Le Mans in 1981 when 200 were sold.

There was a further sale by 'Domaines' in 1982 when the same Messr. Valdevit from Paris who had been involved in the early days of the French Army Cub, bought 14 tons of military Cub spares which had been sold, together with a load of Willis Jeep spares, to a scrap merchant from a town near Limoges. Messr. Valdevit then bought the Cub parts back from the scrapman and saved them for posterity.

10.16 The French Army Cub today

The French Army Cub was built as an 'all-rounder' and incorporated some stronger than standard parts. Subsequent to the Domaines sales, many of these machines were converted in France, making a very fine pre-65 trials bike. Many probably now exist on the continent in this form, as well as a few in this country. However, only a very few machines have survived in the original French Army guise and with full equipment.

10.17 Cubs in the British Army

The first known British military Cub was T20S 66259, delivered to the Fighting Vehicles Research and Development Establishment (FVRDE) at Chertsey, Surrey, in August 1960, for evaluation and feasibility

hall', but they were actually delivered to either the Central Vehicle Depot at Feltham, Middlesex, or the MT School at Bordon, Dorset. These and four others were the only Cubs supplied to the Army in this country.

The four other British military machines were all T20WD models, the first being T20WD 88658. This particular example had been kept on company service as a development machine and was finally sold to a London dealer in December 1964. It was also the first military model to be fitted with the new timing side main bearing, which had first seen the light of day in the Experimental Department and which was subsequently adopted as standard on all Cubs from February 1962. Of the other three machines, two were invoiced to the War Office in Whitehall and the last went to FVRDE in June 1963.

10.18 T20WD, features used on later models

The forty modified T20T machines delivered in July 1961 were designated T20WD. They were based on the 1961 T20T and were the first to incorporate several important features that later became standard equipment on many other Cub models. They were all put into general military use, but six machines were ridden in the forthcoming October ISDT by two army teams, each of three riders.

The machines used the standard camshaft but were fitted with the interference type sports valve springs first

Fig. 10.17b The 1961 Military T20T from the timing side. (The Mike Estall Collection)

studies. The tests were apparently satisfactory because they led directly to the purchase in July 1961 of a further forty machines.

These forty machines, nos. 80921 to 80960, were modified T20T models. The customer for them was shown in the factory records as 'The War Office, White-

Fig. 10.17c An oblique view of the 1961 British Army model. (The Mike Estall Collection)

used on the T20S. The oil pump was a new type with the body in cast iron, giving - it was said - better performance at higher temperatures. The carburettor was a $^{25}/_{32}$in choke Monobloc.

Both footrests were strengthened and there was a new rear chainguard with an enlarged mudshield giving extra protection to the lower chain run. The wheel bearings were sealed, the brake linings were waterproof and the wheels were laced with heavy gauge spokes. The rear chain and sprocket sizes were increased from $^3/_{16}$in to $^5/_{16}$in width but still of $^1/_2$in pitch.

The machine used a full lighting set with stoplight and reflector and had a speedometer. Unusually, considering its intended use, there was a low-level exhaust system which had a clamp around the pipe attached by a bracket to the front engine mounting. This clamp was not a feature carried forward to other models. A twinseat was used and the machines were very smartly turned out in a dark gloss green colour known as Deep Bronze Green, BSC 224.

This model was the first to be fitted with the extended centre stand later used on some sports Cubs. It used the same forged legs as on the T20, but its foot - instead of being a pressing as on the roadster - became a curved tube with small footpads at each end.

10.19 T20WD testing

The first of the forty T20T machines was taken to the MIRA (Motor Industries Research Association) track in Leicestershire on 29th June 1961 for performance testing. Present were John Beckett from the IFVME (Inspectorate of Fighting Vehicles and Mechanical Equipment), Eric Lloyd from FVRDE and Stan Trowell from Meriden. The rider was none other than Triumph tester Percy Tait. No figures are now available from the test, but it was stated that the machine out-performed the TRW 500cc side-valve twin in every respect.

At one point (and it's not known exactly when) the last of these forty machines was sent to IFVME for testing. Some of the ISDT machines apparently lost their petrol tank bolts during the event, allowing the frame to flex and eventually break. For this reason the use of fitted (i.e. very close tolerance) engine and fuel tank bolts was recommended and additional gussets were later fitted under the headstock of some models.

The Inspectorate report for the test machine declared: 'Forks solid. Headraces tight. Noisy timing gears. Front brake noisy. Rear chain too tight. Lighting harness trapped under tank. Horn and cut-out buttons loose'. It was recommended that both the frame and forks be replaced on this machine.

10.20 T20WD: purchase by other armed forces

Henry Vale, long-time head of the Meriden Competition Department, was sent to Denmark to demonstrate a military Cub for possible sale to the Danish armed forces. The year of this adventure is not known for certain, but is thought to have been between 1954 and 1956. In any case his mission was not successful, nothing came of it.

Apart from the French Army Cub, sales of the military models were not exactly a roaring success. A few other countries bought the odd example, and in 1964 three T20WD machines were sent to British Guiana (now Guyana), one went to Hong Kong, one to Jamaica and two to Pakistan. The National Guard in Nicaragua bought three Cubs in September 1965 and two more in March 1966. Apart from the French machines a total of only fifty eight military Cubs were delivered, all originating from Meriden except for the last five examples which came from Small Heath.

10.21 Cubs in the constabulary

Well over two hundred machines, mostly Cubs but also some Terriers, were used by a number of police forces. Several constabularies in the UK, including the Channel Islands and others in France, Pakistan and Kenya, took on small numbers of the Triumph lightweights for patrol and communication duties.

Fig. 10.22a Eight Terriers were delivered to the Chief Engineer, New Scotland Yard, in May 1955. Here they are seen in convoy at the Police Driving School, Aerodrome Road, Hendon. (Courtesy EMAP)

Fig. 10.22b The time? Yes Madam, when the big hand gets to the top it's teatime. (The Mike Estall Collection)

Fig. 10.22c Where beat patrol machines were concerned there was a winner and a loser. The Terrier did not get the job but the 'Noddy Bike' did! (Courtesy EMAP)

10.22 Police Terriers

The very first police machines were a batch of six Terriers sent to the Crown Agent, London, in March 1954 for delivery to the Kenyan police. There they were to be used in the British colonial efforts to control the growing threat from the Mau Mau independence movement.

Then in May 1955 came a delivery of eight Terriers to the Chief Engineer, New Scotland Yard. They were used together with a number of LE Velocettes as part of an efficiency drive enabling the beat officer to cover a far greater patrol area than he could have done on foot.

10.23 Police Cubs in Cambridge

The first Cubs put into police use were sent to Triumph main agent Henry Rose in March, Cambridgeshire, during May 1956, for use by the Isle of Ely police. These machines were fitted with windscreens, panniers and leg shields. Interestingly, although the accompanying photograph clearly shows five machines standing outside the shop premises, the factory despatch books show only four 1956 T20 machines as being delivered to H. Rose.

10.24 The RUC

By far the largest police user was the Royal Ulster Constabulary in Belfast, Northern Ireland, and deliveries totalling eighty-nine machines were made in 1957, 1962 and 1963. The 1962/3 machines were usually fitted with radio equipment consisting of a power pack and aerial mounted on a plywood plate affixed to the stepped rear half of the dualseat. They also had a more powerful alternator, larger battery and amended wiring to hook up the radio equipment.

The petrol tank had four threaded holes set into the top surface for the screws and distance pieces to carry a telephone-type handset, and a conical loudspeaker was mounted on the left handlebar. All the machines were supplied to McIntyre Brothers, May St., Belfast, and were maintained in the police workshops at Castlereagh.

10.25 The T20P

All the earlier police machines were described in the records as 'T20', and it was not until November 1964 that the T20P designation appeared for the police models. By this time, however, Cub production was being transferred to Small Heath and, in fact, only a handful of machines with this model designation were built.

Despatch records show a total of only sixteen T20P machines, the first being delivered in November 1964 and the last in September 1965. All of them went to either Warwickshire or Coventry police, with the possible exception of four machines in November 1964. Here, the records have not been completed and the recipient is unknown.

10.26 T20P livery and equipment

The normal colours for the T20 in 1964/65 were Hi Fi Scarlet and Silver Sheen but, perhaps to maintain tradition and an image of sobriety, all the T20P models were painted black and Silver Sheen. They carried much the same equipment as the earlier RUC models, but in addition the centre stands were strengthened and fitted with a rubber stop. There was also a new large body silencer with a mute, making the machine much quieter than its predecessors.

Fig. 10.23 1956 Cubs lined up outside the premises of Henry Rose, March, Cambridgeshire, May 1956. The machines were used by the Isle of Ely Constabulary. (The Mike Estall Collection)

10.27 Police Super and Mountain Cubs

The final police machine deliveries took place, according to the despatch books, in November 1965 when ten T20B Super Cubs were sent to Paris for the French police, and in November 1967 when one T20SM Mountain Cub was lent to a Sgt. Hughes for use in patrolling a veteran car run.

The French machines were painted black and had a shortened seat for radio equipment, but the real point of interest is the delivery date of November 1965, almost exactly one year before the new Super Cub was announced. This anomoly, the serial numbers of these machines (around T20B 9650) and the delivery dates of other machines despatched with similar serial numbers (October 1968 to July 1969), clearly show that the date in the despatch book must be incorrect.

10.28 Police machines - summary

Altogether the factory despatch records give the delivery details of 14 Terriers and 159 T20 Tiger Cubs sent to the Chief Constables of several major cities or counties, including Oxford, the North Riding of Yorkshire, Preston, Cardiff, Dudley, Birkenhead, Salford, Leeds, Warwickshire and Northamptonshire. Additionally, 12 T20C Competition Cubs were shared between the

Monmouthshire, Cumberland and Westmorland constabularies. There were also 16 T20P models, 11 T20B Super Cubs and 1 T20SM Mountain Cub, making a total of 213 machines delivered for police service.

Fig. 10.24a T20 72269 Police 'demonstration' machine outside the RUC Headquarters, Belfast, 1962. (The Mike Estall Collection)

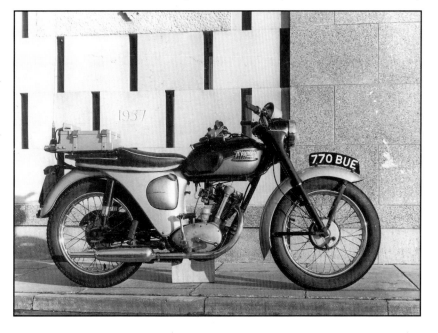

Fig. 10.24b Belfast PC Harry Vaughan making a call. (The Mike Estall Collection)

Fig. 10.24d The demonstration machine at a local event - stripped of its radio equipment! (The Mike Estall Collection)

Fig. 10.24c A 1963 Police demonstration machine in non-standard Black and Silver. Note the radio pack and handset; also the enlarged silencer. (The Mike Estall Collection)

Military, Police and the Utilities

Fig. 10.24e The same police model, registered 858 FAC, kitted out with fairing, legshields and panniers. (The Mike Estall Collection)

Fig. 10.24f 1963 T20 machines destined for the RUC being loaded onto a Vickers Vanguard. Note the reinforced centre stands, larger silencers and extensive use of corrugated cardboard. This is part of an order for twenty seven machines, delivered to McIntyre Brothers, Belfast, on 30th January 1963. (The Mike Estall Collection)

Fig. 10.27 One of the ten French police Super Cubs. According to the despatch book these machines were delivered one year before the model was announced. (Courtesy Jean Caillou, Paris, France)

10.29 Local authority users

In this country little use was made of the Cub by local authorities or public utility companies. One T20 went to Worthing Corporation; four were sent to the Fife County Council and one was used by the Wisbech District Water Board, all in 1957. The City Engineer in Coventry had two T20 machines in 1960 and Woolwich Town Council bought a Super Cub in 1967. And that was it!

A far greater number of machines were sent overseas, but directly to only two customers. The first foreign delivery of this type was of two T20C machines to the Medical Research Council in Gambia. By far the largest number were 208 Bantam and Super Cubs sent to Thailand between July 1966 and 1967 for use by the Metropolitan Electricity Authority in Bangkok. The records do not show how many of each model there were,

and the duties to which these machines were put is unknown.

A third user was the fire service in Paris, France, which had three Super Cubs supplied by CGCIM. These machines were painted in normal civilian colours.

10.30 Summary

The institutional use of the Terrier and Cub cannot really be described as a success story, despite the quite large number of machines sold to this type of user. Having said that, the three major customers in this area - the Royal Ulster Constabulary, the Bangkok Metropolitan Electricity Authority and, of course, the French Army - no doubt received sterling service from their mounts because they kept coming back for more! Apart from these three customers there were few others.

Chapter 11
In memoriam

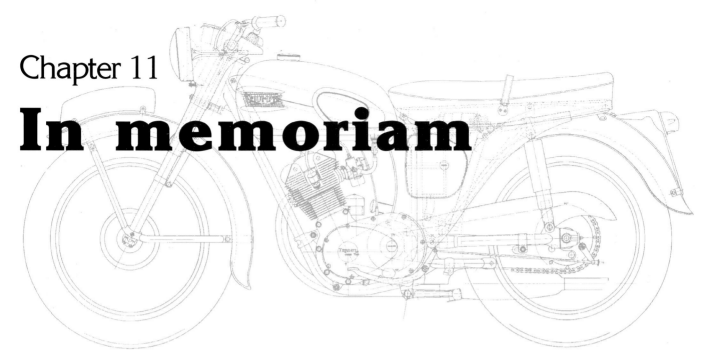

11.1 Early hopes

The motor cycle enthusiast, not long after the end of the second world war and in a period of belt-tightening austerity, welcomed with open arms the arrival of the Terrier. Here was a brand new design, a very good-looking machine from a major manufacturer, giving outstanding all-round performance. The advent of the Tiger Cub one year later was even more welcome and provided yet another alternative to an ocean of sometimes rather mediocre two-strokes. Both models had the charisma of being Triumph machines and were in great demand.

Early problems with delivery soon cleared and sales began in earnest - but so did the occurrence of other problems, such as those mentioned in chapter 3. Slowly, over a period of years, these difficulties were overcome and the machines continued to sell well. Edward Turner's principle of making the minimum amount of metal do the maximum amount of work had been the direct or indirect cause of most of the trouble. It was a flawed philosophy which sounded good on paper but in reality had been the cause of much heartache. To put it simply, the machine had too much performance for its own good and was for a long time not strong enough to live up to its rider's expectations.

However, many thousands of riders learned to ride on a Cub and it was extensively used as a commuter machine. It had excellent acceleration, was fast for its size, inexpensive, light in weight and easily handled. It was also, thankfully, easy to work on and spares were cheap. Yet, despite its reputation for unreliability and break-down (often severely testing marque and model loyalty), people loved the Cub and kept coming back for more.

11.2 Trouble on the horizon

Fashions in transport were changing. By the start of the sixties, post-war austerity had ended, and the 'Never had it so good' days had arrived; wages were going up and cars were becoming more plentiful. The arrival of Alec Isigonis' Mini, and other small cars, meant that commuter transport requirements were just as likely to be satisfied by four wheels as by two. Britain's road system was also being developed and the family car, rather than Dad's motorbike, increasingly became the usual mode of transport. As an everyday vehicle, the days of the motor cycle were becoming numbered.

From 1957 sales of the various Cub models had increased every year, finally peaking at over 13,000 machines for the 1960 season. This figure was no doubt boosted by British legislation which restricted learner riders to machines of less than 250cc capacity.

The 1961 season was pretty good too, with over 12,000 Cubs sold, but then the decline really set in and only half that number were sold in 1962 and 1963, and even less in 1964. By the time production had been turned over to BSA in early 1965, the Japanese lightweights were having a profound influence on sales figures. Sales in the remaining four years totalled only about 14,000 machines and the model was decidedly moribund. Deliveries continued at a dribble into 1969 and 1970, but these machines were mostly unsold examples taken out of the warehouse stockpiles, often at clearance prices.

Oriental models were biting deeply into the sales of all home manufacturers, and the sheer volume of machines produced by the Japanese factories gave them a good price advantage over the Cub. Their models were also better equipped, faster, more powerful, quieter, more fashionable and did not leave oil stains all over the place.

11.3 The new Japanese industry

The writing had been on the wall for years, but many

people in positions of influence had been either unwilling or unable to see it. As long ago as 1960, after visiting Japan, Edward Turner had warned everyone about what was happening there. By 1965 all the major improvements made to the Cub over the years resulted in a very tough little engine, but it was all too late as the machine was by now an obsolete design, outshone and outperformed by its Oriental and Continental competitors, and within a very few years it was gone.

The Japanese had the advantage of brand new factories, plant and equipment paid for by the west after the war, and also had a clean design slate. They were unencumbered by long traditions in motor cycle design or manufacture, and in Turner's words were 'Highly intelligent, very energetic, purposeful people, all geared to an economic machine with an avowed object of becoming great again ...' They were capable of producing machines in large numbers and of a very high quality which were economical to buy and run, were very quiet, in many cases had electric starters, and did not leak oil.

Turner had prophetically warned of the threat after his Far-Eastern visit but no one took any notice; after all, sales were booming here and British manufacturers could sell every bike they built. But just to put things into perspective, in 1960 the Japanese motor cycle industry produced nearly four times the number of machines that we did, and that was the peak year for UK production! After that, they made even more and we made fewer.

Of course, many of these Japanese machines were small scooters and mopeds but the production capacity was in place, the workforce was highly motivated and it was only a matter of time (in fact, a very short time) before Britain's motor cycle industry was forced into terminal decline by the advent of direct competitors to our larger capacity models.

It had always been the larger machines that made the profits which sustained the industry as a whole, underwriting production of smaller models. Once those profits went into decline there was no hope for the lightweights. They could not compete in the marketplace by themselves against a Japanese industry that was already manufacturing small capacity machines by the million.

11.4 Sales to the end

Throughout its early life the Cub and variants were competitively priced, often slightly less than other two-stroke machines of the same capacity and always considerably less than most other four-strokes of 250cc. This situation prevailed until about 1960, by which time the Japanese machines were beginning to show their mettle.

In the USA, for example, in June 1960, a twin cylinder overhead cam, 305cc electric start Honda could be bought for $595, only $50 more than a standard T20 and less than a T20S sports Cub. By 1962 the American sporting rider could have spent $683 on a T20SC 15bhp Tiger Cub - or $639 on a 250cc 25bhp Parilla Trailmaster - or $690 on a 250cc 24bhp twin cylinder Honda Scrambler. Which would you have bought?

11.5 Cub deliveries abroad

From Aden and Argentina to Zanzibar and Zimbabwe, the Terrier and Cub were delivered to every continent. There were 24 different model designations (plus some others with 150cc engines) and they were delivered to 153 different countries in the eighteen years between July 1953 and June 1970. They went as far north as Iceland, south to the Falklands, east to Japan and west to the Pacific shores of the USA.

They were delivered to some unlikely recipients too, such as His Majesty the King of Nepal, who had several Cubs and a number of larger machines as well. European royalty was not without interest in the Meriden lightweights either. Baudouin, King of the Belgians, had a private collection of Tiger Cubs imported through the Belgian main Triumph distributor M. Joseph Decat, with whom he used to go riding in the royal grounds.

There were other countries where only a handful of machines were sent. For example, Ascension Island, Bolivia, Bulgaria, Faroe Islands, Guadeloupe, Java, Liberia, Libya, Montserrat, Niger, Poland, Reunion Island, Rumania, Tahiti, SW Africa and St. Helena. A rummage through the despatch books makes fascinating reading and a close scrutiny of Appendix 6, where these books are summarized, will reveal some interesting facts concerning delivery of these machines.

Some of the larger overseas sales are shown in the accompanying table. Also shown under the "Per 1000 Pop'n" heading are the total deliveries for all years divided by the number of thousands of the country's population figure in 1964. Aden, for example, had 122 machines and the population was 220,000. Therefore, 122 divided by 220 equals 0.55. In other words, the larger the figure, the greater the Terrier and Cub density per thousand population.

11.6 The end of the line

By March 1968 the end was in sight. General Tiger Cub production stopped in October 1968, although there were 248 examples of the 'T20 Bantam 175' built in early 1969 and one final batch of 53 Super Cubs built at the end of June 1969. The last machines delivered were some French Army models in May 1970 and the very last delivery of all was a Super Cub, to Slocombes of Neasden, London, on 18th June 1970.

The factory records show altogether the delivery of 112,672 Terriers and Cubs of all models, from July 1953 to June 1970. To put this figure into some sort of perspective there were approximately 243,000 twin cylinder models delivered from Meriden in the same period and roughly 20,000 scooters.

The decline of the British motor cycle industry has been covered in depth by other writers, but suffice to say that, by the mid-sixties, the Cub - which had been such an exciting and innovative design little more than a decade earlier - had become hopelessly outdated. All the basic flaws in the engine design had been gradually corrected until by the time they had just about got it right, they decided to stop making it! This irony will not be lost on the reader. By 1968, when Cub production had

Country	Bikes		Pop (000s)		Per 1000 Pop'n	Comments
Aden	122	÷	220	=	0.55	
Angola	604	÷	4833	=	0.12	
Australia	1313	÷	10916	=	0.12	737 machines in 1954/5 is more than half the Australian total
Belgium	599	÷	9290	=	0.06	
Bermuda	1049	÷	47	=	22.32	The winner! The highest proportion delivered for any country to which significan numbers of machines were sent
Burma	229	÷	23664	=	0.01	225 T20 sent in 1961
Canada	813	÷	18928	=	0.04	
Ceylon	993	÷	10625	=	0.09	
Denmark	664	÷	4654	=	0.14	Nearly 500 T20 delivered in 1955/6
Eire	1460	÷	2841	=	0.51	Many machines delivered unassembled
Finland	554	÷	4562	=	0.12	212 T20 models delivered in 1961/2
France	1654	÷	48090	=	0.03	Includes all military machines
Guatemala	290	÷	4095	=	0.07	114 T20 delivered in 1959
Holland	146	÷	12079	=	0.01	The majority of machines were delivered in 1954/5
Hong Kong	521	÷	3600	=	0.14	The majority of deliveries were made after 1964
India	388	÷	449381	=	0.001	Deliveries steady until 1957 when legislation reduced imports
Iran	713	÷	22357	=	0.03	281 Bantam and Super Cubs to Tabarrok Brothers in Teheran
Jamaica	455	÷	1706	=	0.27	Over 200 Super Cubs sent
Kenya	454	÷	8847	=	0.05	
Malaya/ Malaysia	1819	÷	10364	=	0.17	
Malta	150	÷	326	=	0.46	
New Z'land	686	÷	2415	=	0.28	
Nicaragua	604	÷	1524	=	0.40	
Nigeria	1094	÷	55654	=	0.02	777 T20 delivered in 1959/61
Norway	497	÷	3681	=	0.14	Few deliveries after 1962
Pakistan	2570	÷	98612	=	0.03	2,265 Cubs sent to Atherton Brothers in Karachi or Lahore
Singapore	1439	÷	1799	=	0.80	
South Africa	893	÷	17057	=	0.05	
Sweden	204	÷	7627	=	0.03	177 Terriers delivered in 1954
Tanganyika	365	÷	362	=	1.01	
Thailand	603	÷	29700	=	0.02	Orders for 208 T20B by the Bangkok Electricity Board
Uganda	417	÷	7016	=	0.06	
U. K.	61428	÷	53812	=	1.14	Approximately half the total production was delivered within the U.K.
U.S.A.	21453	÷	191637	=	0.11	Nearly one fifth of the total production went to the U.S.A.
All countries				=	0.059	Delivered to 153 countries and 1,911,044,000 population

all but ended, it was outclassed, outperformed and outpriced by Oriental machinery and it was definitely time to go.

The 1969 catalogue was devoid of a T20 model for the first time in many years, and the new TR25W Trophy model had been announced for the 1969 season as the replacement for the Tiger Cub. However, the basic design of the engine unit survived, starting with the C15 in 1957, continuing through to the B40, badge-engineered 1969 B25 and TR25W models, and several other derivatives whose mechanical lineage could be traced right back to the Terrier.

11.7 The Tiger Cub today

In many ways the Triumph lightweights are just as popular today as they were three or four decades ago, and the exhaust crackle of a Tiger Cub can still be heard every now and then. They say that distance lends enchantment to the view and the Cub is still regarded by many as 'a great little bike!' On the other hand, it has to be said, too, that there are many others who, even with the strongest rose-tinted glasses, cannot regard the model as having been anything but 'a load of trouble.'

It can be said that both views have some validity, but these days things have changed more than somewhat. Modern oils are so much better than those of forty years ago. Electronics can now replace many of the older and less reliable electrical components, and their installation also considerably simplifies the wiring.

However, the biggest factor lies with the owners themselves. A large proportion of riders then were inexperienced youngsters whose sole desire was to ride the bike fast, usually without the benefit of a warm-up period, and make as much noise as possible. Routine maintenance, which was frequently ignored, all too often, became breakdown maintenance and the gentle art of doing up nuts and bolts correctly was a skill yet to be acquired by the average teenager. As often as not the result of this neglect was a short life for his machine, further reinforcing the Cub's reputation for lack of mechanical stamina.

Nowadays those same owners are considerably older, wiser and a great deal more responsible. They will probably have many years' mechanical experience to call upon and will treat their mounts with the respect and consideration they deserve. They have also perhaps gone through a long period raising a family and now, having some leisure time, would like to rebuild and ride the machine they owned in their youth.

Not only that but the object of their attentions will have the advantage of being, in monetary terms, an appreciating asset. What better reason to give the wife, who sees hard earned family money being diverted to the rusty wreck in her husband's workshop? However, the born-again enthusiast is unlikely to point out to his spouse that the cost of the project will probably far exceed its inherent value! With small machines like the Cub, restoration is a labour of love constrained only by available financial resources. The Tiger Cub is still regarded with great affection and many derelict examples are now being pulled from old sheds into the light of day to be lavished with far more money, care and attention than they ever had in the past.

Tiger Cubs, and the occasional Terrier, are now entered in many vintage motor cycle rallies up and down the country. Go to any Pre-65 trial and you will see several examples churning their way through the mud or over the rocks. Some are even used today for their original purpose as a commuter machine!

The Tiger Cub and Terrier Register currently records the survival of roughly two and a half thousand complete machines and about the same number of frames and engines. Of those machines a large proportion still retain their original matching frame and engine numbers. Some of these have been in the same family for two, or even three, generations, being faithfully passed down from father to son. If that isn't a sign of real affection, then what is?

One of the great assets of the Cub and Terrier has been interchangeability of parts. Any Cub or Terrier engine will fit into any Cub or Terrier frame and many other mechanical and cycle parts can also be mixed and matched. This means that the bike found at the back of some chicken shed, having lain there since it was abandoned and forgotten by its owner years ago, often has a different engine from the original, the wrong petrol tank, odd sized wheels and a cocktail of other miscellaneous bits and pieces.

But here the Tiger Cub Register can sometimes come to the rescue by perhaps locating the engine that matches the frame, or vice-versa. The author's information service also does much to ensure that the correct parts are used to bring the machine back to its former glory, helping the owner avoid spending his money buying the wrong parts or painting them the wrong colour.

There is a great sense of satisfaction and achievement to be had in converting a rusty wreck into something that looks as if there has been a time warp and it has just been ridden past the factory gates. Total strangers approach in the street, saying 'I had one of those when I was a lad', and the owner has made a new friend. Many machines are now going through this process and are, at the same time, giving a great deal of pleasure to everyone involved.

To paraphrase an ancient incantation 'The Cub is dead. Long live the Cub'.

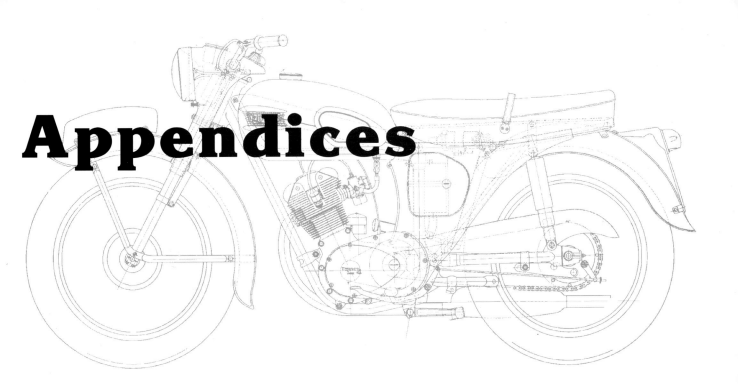

Appendices

A 1954 T15 Terrier from the drive side. The colour was Amaranth Red, including wheel rim centres which were lined in gold. (Courtesy Jim Davies)

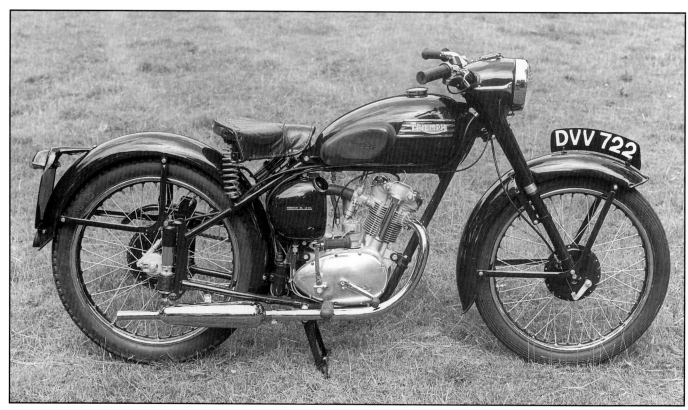

The 1954 T15 from the timing side. This machine has been restored by the author and also converted to 125cc. (Courtesy Jim Davies)

Appendix 1.1 Model profile

T15 Terrier - 1954 to 1956

Machines T15 101 (July 1953) - T15 31463 (April 1957).

Bore & stroke:	57mm (2.244in) x 58.5mm (2.303in).				
Compression ratio:	7:1.				
Camshaft:	Standard. Valve clearances - both 0.010in (cold).				
Ignition type:	Lucas AC/DC with battery. RM 13 alternator.				
Ignition timing:	8° BTDC (= 0.4mm or $^1/_{64}$in), static. (= 32° fully advanced).				
Points gap:	0.014in - 0.016in.				
Spark plug:	Champion L7 or L10S, Lodge HN or H14, KLG F70 or F80. Gap 0.025in.				
Carburettor:	Amal 332/1 or 332/2, $^{11}/_{16}$in choke size.				
Gearbox ratios:	Standard ratios.	1st	3rd	2nd	4th
	Mainshaft -	16T	25T	20T	28T
	Layshaft -	29T	20T	25T	17T
Wheel & tyre sizes:	Front	2.75 x 19, on WM1 rim.			
	Rear	2.75 x 19, on WM1 rim.			
	Tyres	Dunlop Universal, block pattern.			

Sprockets & chains: Engine 19T.   Clutch 48T.   Primary chain $^3/_8$in x $^7/_{32}$in x 62 links, simplex.
Gearbox 17T.   Rear wheel 48T.   Rear chain $^1/_2$in x $^3/_{16}$in x 112 links.

Capacities:	Petrol.	2$^5/_8$ gallons (= 11.92 litres). 'Teardrop' shape tank.
	Oil (1954/5).	2$^1/_4$ pints (= 1.28 litres).
	(1956).	2$^3/_4$ pints (= 1.57 litres).
	Gearbox.	$^1/_3$ pint (= 200cc).
	Primary chain.	$^1/_2$ pint (= 300cc).
	Suspension.	Grease, as required, front and rear.

Exhaust system: Low level as standard. High level option - UK in 1955, USA in 1956.

Colours:
All markets for all years, except USA 1955.
All painted parts, except number plates -   Amaranth Red.
Wheel rim centres -   Amaranth Red lined in gold.
Tank badges -   White lettering on an Amaranth Red and chrome backing strip.

USA for 1955.
All painted parts -   Black.
Wheel rim centres -   Black lined in gold.
Tank badges -   White lettering on black and chrome backing strip.
Front brakeplate -   Chrome plating.
Both mudguards were plain section - without a raised central rib. (T20 mudguards were both ribbed).

Saddle or twinseat: Twinseat initially all black, then from no. 3001 white piping was added around the top edge.
(The seat and pan were the same as the sprung wheel twins but with different fittings).
Saddle had a black cover and parallel sided springs. (A maroon cover has also been seen).

Notes:
Many details changed early on as experience was gained. Initially a roller big end was used. This was changed to a plain type at no. 3905.
A sludge trap was put in the TS flywheel from no. 4859.
New big end and TS mains bush material (Vandervel VP3) from no. 24090.
Thicker rimmed flywheels and a new crankpin with flywheels spigot enlarged from $^3/_4$in to $^{13}/_{16}$in at no. 18597.
Cylinder barrel was the 'Round' type and there were no fins on the cylinder head inlet rocker box.
Oil pump feed plunger diameter increased from $^3/_{16}$in to $^1/_4$in from no. 3905. Auxilliary balls from no. 10197.
There were several changes to the clutch, at no. 3905, no. 5360, no. 8022, no. 8165, no. 8259, no. 11621, no. 16515, no. 17389.
Primary chain filler cap – up to no. 3904 - top of inner DS cover, sited above the alternator from no. 3905, then at the clutch centre from no. 8518.
The centre stand lug was strengthened from no. 2278. The tubular centre stand was replaced by a forged type at no. 6670.
Frames were fitted with a lug for the optional prop stand, probably from no. 3001.
New handlebar bend for 1956 and longer levers. At the same time cut-outs were made in the nacelle top for the clutch and brake cables.
Reinforcing struts were put inside the petrol tank from no. 17389.
Rear light - Lucas L529, but L525 used on some 1954 machines.
HT coil - up to no. 3905, mounted under the seat, thereafter on the rear mudguard, behind the gearbox.
Rectifier - 1954 - Lucas 47103A = 4$^1/_2$in diameter. 1955/6 - Lucas 47111A = 2$^3/_4$in diameter.

Some history: The Terrier first appeared at the 1952 Earls Court Show but the first deliveries were not made until late July 1953. The factory year ran from 1st August to the following 31st July, and so, to all intents and purposes, there were no 1953 Terriers. (There was only one machine actually delivered before the end on July 1953). With the exception of this machine all Terriers despatched from 1.8.53 were 1954 models.
Production of the Terrier ceased at the end of the 1956 year although a few were sent to Indonesia in April 1957.
For road test figures see Appendix 7.

The serial numbers of machines in each year were: (n.b. from no. 3000 Cubs were listed in the same number series as Terriers).
1954 machines   no. 101 to no. 8517.   (Parts List No.1 covers no. 101 – 3904).
    (Parts List No.2 covers no. 3905 onwards).
1955 machines   no. 8518 to no. 17388.   (Parts List No.3 covers all machines up to no. 17388).
1956 machines   no. 17389 to no. 26275.   (Parts List No.4 covers all machines from no. 17389).

Machines built: 1954 = 6,240   1955 = 2,074   1956 = 893   1957 = 30   Total = 9,237.

Analysis of sales (all years): UK = 50.0%.   USA = 15.7%.   Others = 34.3%.

A 1954 T20. A high level exhaust was standard for the 1954 T20. (The Mike Estall Collection)

A 1955 T20. The standard T20 exhaust has now become a low level system. (The Mike Estall Collection)

Appendix 1.2 Model profile

T20 Tiger Cub - 1954 & 1955

Machines T20 3000 (March 1954) - T20 17388 (July 1955) (except for no. 16881 - 16982 which were all built to the 1956 pattern).

Bore & Stroke:	63mm (2.480in) x 64mm (2.520in).				
Compression ratio:	7:1				
Camshaft:	Standard. Valve clearances - both 0.010in, (cold).				
Ignition type:	Lucas AC/DC with battery. RM 13 alternator.				
Ignition timing:	8° BTDC (= 0.4mm or ¹/₆₄in), static (= 32° fully advanced).				
Points gap:	0.014in - 0.016in.				
Spark plug:	Champion L7 or L10S, Lodge HN or H14, KLG F70 or F80. Gap 0.025in.				
Carburettor:	Amal 332/1, 332/2 or 332/3, ³/₄in choke size.				
Gearbox ratios:	Standard ratios.	1st	3rd	2nd	4th
	Mainshaft -	16T	25T	20T	28T
	Layshaft -	29T	20T	25T	17T
Wheel & Tyre sizes:	Front.	3.00 x 19, on WM1 rim.			
	Rear.	3.00 x 19, on WM1 rim.			
	Tyres.	Dunlop Universal, block pattern.			
Sprockets & Chains:	Engine19T.	Clutch 48T.	Primary chain ³/₈in x ⁷/₃₂in x 62 links, simplex.		
	Gearbox 18T.	Rear wheel 48T.	Rear chain ¹/₂in x ³/₁₆in x 112 links.		
Capacities:	Petrol.	2⁵/₈ gallons (= 11.92 litres). 'Teardrop' shape tank.			
	Oil.	2³/₄ pints (= 1.28 litres).			
	Gearbox.	¹/₃ pint (= 200cc).			
	Primary Chain.	¹/₂ pint (= 300cc).			
	Suspension.	Grease, as required, front and rear.			

Exhaust system: 1954 - High level, with low level option.
1955 - Low level, with high level option.

Colours:
	All markets -	Shell Blue Sheen and black.
	Petrol tank -	Shell Blue.
	Mudguards -	Shell Blue with a black centre stripe, lined in white.
	Wheel rim centres -	Shell Blue lined in black.
	Front Brakeplate -	Shell Blue. (USA - chrome plated in 1955).
	Tank badges -	White lettering on black and chrome backing.
	Other painted parts -	Black.
	Both mudguards were ribbed in the centre.	

Saddle or twinseat: Twinseat all black with white piping around the top.
(The seat and pan were the same as used on the sprung wheel twin cylinder models).
Saddle had parallel sided springs.

Notes:
Cylinder barrel - 'Round' shape. No finning on the cylinder head inlet rocker box.
Plain big end from the start. 'Thin' type flywheels, with a sludge trap in TS flywheel from no. 4859.
There were several changes in the clutch, at no. 3905, no. 5360, no. 8022, no. 8165, no. 11621 & no. 16515.
Primary chain filler cap - none up to no. 3904, above the alternator from no. 3905, then at the clutch centre from no. 8518.
Centre stand - tubular in section for 1954. New forged stand from no. 6670.
New nacelle top with cut-outs for the clutch and brake cables from no. 17258.
Prop stand option, (and lug fitted to frame), probably from the start.
HT Coil - up to no. 3905, under the seat. Thereafter on the rear mudguard, behind the gearbox.
Rectifier - 1954 - Lucas 47103A = 4¹/₂in diameter. 1955/6 - Lucas 47111A = 2³/₄in diameter.
Rear light unit - Lucas L529.

Some history: The Tiger Cub was developed from the T15 Terrier introduced a year earlier. Apart from the colour, bore and stroke, overall gearing, mudguard type and tyre sizes, they were to all intents and purposes the same.
The first Cub shown in the factory despatch books was T20 3000 and was possibly the machine exhibited at the 1953 Earls Court Show. The first production Cub was despatched from the factory in March 1954. Thereafter, and until the Terrier's demise, Cubs and Terriers were manufactured together and shared a common numbering sequence.
The production year ran from 1st August to the following 31st July. Frame and engine numbers were always the same.
For road test figures see Appendix 7.

The serial numbers of machines in each year were:
1954 machines no. 3000 and no. 3002 to no. 8517. (Apart from T20 3000 all machines from no. 101 - no. 3001 were Terriers).
1955 machines no. 8518 to no. 17388. (n.b. 102 1955 machines were built for Sweden to the 1956 specification).
Parts List No.3 covers all 1954/5 Cubs.

Machines built: 1954 = 1,974   1955 = 6,446

Analysis of sales (both years): UK = 62.8%.   USA = 16.8%.   Others = 20.4%.

A 1956 T20 'Chubby Cub' restored by the author. The 16 inch wheels gave a very low seat height. The ignition coil cover and the chainguard were both unique to this year and model. (The Mike Estall Collection)

Appendix 1.3 Model profile

T20 Tiger Cub - 1956

Machines T20 17389 (August 1955) - T20 26275 (July 1956).
(Plus 1955 machines T20 16881 - 16982 which were all built to the 1956 pattern).

Bore & Stroke:	63mm (2.480in) x 64mm (2.520in).
Compression ratio:	7:1.
Camshaft:	Standard. Valve clearances - both 0.010in, (cold).
Ignition type:	Lucas AC/DC with battery. RM 13 alternator.
Ignition timing:	4° BTDC, static. (= 28° fully advanced).
Points gap:	0.014in - 0.016in.
Spark plug:	Champion L7 or L10S, Lodge HN or H14, KLG F70 or F80. Gap 0.025in.
Carburettor:	Amal 332/3, $^3/_4$in choke size.

Gearbox ratios:	Standard ratios.	1st	3rd	2nd	4th
	Mainshaft -	16T	25T	20T	28T
	Layshaft -	29T	20T	25T	17T

Wheel & Tyre sizes:	Front.	3.25 x 16, on WM2 rim.
	Rear.	3.25 x 16, on WM2 rim.
	Tyres.	Dunlop Universal, block pattern.

Sprockets & Chains:	Engine 18T.	Clutch 36T.	Primary chain $^1/_2$in x $^3/_{16}$in x 48 links, simplex.
	Gearbox 17T.	Rear wheel 54T.	Rear chain $^1/_2$in x $^3/_{16}$in x 116 links.

Capacities:	Petrol.	3$^1/_8$ galls (= 14.0 litres). New 'Flat' shaped tank.
	Oil.	2$^3/_4$ pints (= 1.56 litres).
	Gearbox.	$^1/_3$ pint (= 200cc).
	Primary Chain.	$^1/_2$ pint (= 300cc).
	Suspension.	Grease, as required, front and rear.

Exhaust system:	Low level, but with a high level option in the USA.

Colours:	All markets -	Shell Blue Sheen and black.
	Petrol tank -	Shell Blue.
	Mudguards -	Shell Blue with a black centre stripe, lined in white.
	Wheel rim centres -	Shell Blue lined in black. (Some late 1956 machines had unpainted rims).
	Front Brakeplate -	Shell Blue. Chrome plating option for the USA.
	Tank badges -	White lettering on black and chrome backing.
	Other painted parts -	Black.
	Both mudguards were ribbed in the centre.	

Twinseat:	All black with white piping around the top. (Seat and pan were the same as on the sprung wheel twins).
	(The saddle option was dropped at the end of 1955).

Notes:
Machines no. 17389 to no. 18596 used flywheels with a thicker outer rim but still retaining the $^3/_4$in shouldered crankpin.
At no. 18597 the shoulder diameter of the crankpin into the flywheels was increased to $^{13}/_{16}$in.
From engine no. 24090 a new VP3 conrod bearing was fitted using the same material as the timing side main bearing.
From engine no. 24244 a new drive side inner cover was used, with greater chain clearance. This also had threads incorporated in two of the screw holes to facilitate removal of the cover by means of a new special tool. New, heavier crankcase outer covers.
A new primary chain was used - now the same size as the rear chain, and there was a new clutch bronze bearing ring with 16 x $^5/_{32}$in balls.
The petrol tank changed for 1956 to a broader, flatter shape with new badge backing strips having a different curvature to allow for the new tank profile. There was a new oil tank with increased capacity and with its cap now placed in the middle of the tank top instead of being sited at the front corner, as in former years.
The wheel size was reduced to 16in.
Frame modified to give additional clearance for the rear tyre. New (shorter) centre and prop stands.
New handlebar bend with longer levers.
There was a new design of gear indicator from machine no. 26106.
The new small diameter wheels lowered the dualseat height by 1$^1/_2$in to 28$^1/_2$in.
There were three items with a design unique to this year and model:
    an underseat plastic valance protecting the HT coil and rectifier from the elements
    a rear chainguard of much fuller design and an oval hole for access to the rear suspension
    a rear mudguard with a fixed bracket for attachment to the frame.

For road test figures see Appendix 8.

The production year ran from 1st August 1955 to 31st July 1956. Engine and frame numbers were always the same.

Parts List No.4 covers all 1956 machines, both T15 and T20 models.

The number of T20 Cubs built in 1956 = 7,792.

Analysis of sales: UK = 59.1%.   USA = 12.8%.   Others = 28.1%.

The 1957 T20 with the new swinging fork frame - from the drive side. Another machine restored by the author. (The Mike Estall Collection)

The 1957 T20 - from the timing side. (The Mike Estall Collection)

Appendix 1.4 Model profile

T20 Tiger Cub - 1957

Machines T20 26276 (August 1956) - T20 35846 (July 1957).

Bore & Stroke:	63mm (2.480in) x 64mm (2.520in).
Compression ratio:	7:1.
Camshaft:	Standard. Valve clearances - both 0.010in, (cold).
Ignition type:	Lucas AC/DC with battery. RM 13 alternator.
Ignition timing:	4° BTDC, fully retarded. (= 28° fully advanced).
Points gap:	0.014in - 0.016in.
Spark plug:	Champion L7 or L10S, Lodge HN or H14, KLG F70 or F80. Gap 0.025in.
Carburettor:	Amal 332/3 or 332/4, $^3/_4$in choke size.

Gearbox ratios:	Standard ratios.	1st	3rd	2nd	4th
	Mainshaft	16T	25T	20T	28T
	Layshaft	29T	20T	25T	17T

Wheel & Tyre sizes:	Front.	3.25 x 16 on WM2 rim.
	Rear.	3.25 x 16 on WM2 rim.
	Tyres.	Dunlop Universal, block pattern.

Sprockets & Chains:	Engine 18T.	Clutch 36T.	Primary chain $^1/_2$in x $^3/_{16}$in x 48 links, simplex.
	Gearbox 17T.	Rear wheel 54T.	Rear chain $^1/_2$in x $^3/_{16}$in x 116 links.

Capacities:	Petrol.	$3^1/_8$ galls (= 14.0 litres). 'Flat' shaped tank.
	Oil.	$2^3/_4$ pints (= 1.56 litres).
	Gearbox.	$^1/_3$ pint (= 200cc).
	Primary Chain.	$^1/_2$ pint (= 300cc).
	Front fork.	$^1/_8$ pint (= 75cc).

Exhaust system:	Low level. No high level option from now on.

Colours:	All markets -	Crystal Grey and black.
	Petrol Tank -	Crystal Grey.
	Mudguards -	Crystal Grey with black centre stripe, lined in white.
	Front Brakeplate -	Crystal Grey. Chrome plating option in USA.
	Tank Badges -	White lettering on a black and chrome backing.
	Other painted parts -	Black.

Twinseat:	New seat shape. Seat top colour usually grey, black sides with grey lower styling band and piping.

Notes:
New cylinder head with finning on the inlet rocker box. 'Round' barrel design still used.
Vandervel VP3 material used in big end bush, timing side main bearing bush and swinging arm bushes.
Primary chain tolerances tightened. New drive side inner and outer covers.
Taper piston ring used in the middle groove.
New frame incorporating swinging arm rear suspension and a new oil damped front fork to replace the earlier plunger frame and grease filled front suspension.
The fork stanchions are now 0.010in smaller in diameter than the early type and have a rubber oil seal and a new top bush.
New chrome plated oil seal holder. There is a new middle lug and stem (the top lug remains the same).
Centre and prop stands now longer than those for the 1956 model.
New silencer, slightly larger than 1956.
New rear mudguard for the swinging arm frame.
New rear wheel with butted spokes on both sides.
Wheel rim centres no longer painted.
New chainguard, now with separate front and rear parts.
New forged rear brake lever and a new brake rod.

For road test figures see Appendix 8.

The production year ran from 1st August 1956 to 31st July 1957. Engine and frame numbers were always the same.

Parts List No.5 covers all 1957 (and 1958), models.

The number of production T20 Cubs built in 1957 = 7,714.

Analysis of sales: UK = 61.3%.   USA = 12.3%.   Others = 26.4%.

A 1958 T20 picture from the sales brochure. The two visible differences from the 1957 model were the deeply valanced rear mudguard and, from machine no. 38130, the 'Mouth Organ' tank badges. (The Mike Estall Collection)

Appendix 1.5 Model profile

T20 Tiger Cub - 1958

T20J Junior Cub - 1958

Machines T20 35847 (August 1957) - T20 45311 (July 1958).
Including T20J models from T20J 36365 (October 1957).

Bore & Stroke:	63mm (2.480in) x 64mm (2.520in).
Compression ratio:	7:1.
Camshaft:	Standard. Valve clearances - both 0.010in, (cold).
Ignition type:	Lucas AC/DC with battery. RM 13 alternator.
Ignition timing:	4° BTDC, fully retarded. (= 28° fully advanced).
Points gap:	0.014in - 0.016in.
Spark plug:	Champion L7 or L10S, Lodge HN or H14, KLG F70 or F80. Gap 0.025in.
Carburettor:	Amal 332/3, $^3/_4$in choke size. From machine no. 39167 a Zenith 17MXZ-CS5. Restrictor on T20J.

Gearbox ratios:	Standard ratios.	1st	3rd	2nd	4th
	Mainshaft	16T	25T	20T	28T
	Layshaft	29T	20T	25T	17T

Wheel & Tyre sizes:	Front	3.25x 16 on WM2 rim
	Rear	3.25x 16 on WM2 rim
	Tyres	Dunlop universal, block pattern

Sprockets & chains:	Engine 19T.	Clutch 48T.	Primary chain $^3/_8$in x $^7/_{32}$in x 63 links, duplex
	Gearbox 18T.	Rear wheel 46T.	Rear chain $^1/_2$in x $^3/_{16}$in x 112 links

Capacities:	Petrol	$3^1/_8$galls (= 14.0 litres). 'Flat' shaped tank.
	Oil.	$2^3/_4$ pints (= 1.56 litres).
	Gearbox.	$^1/_3$ pint (= 200cc).
	Primary Chain.	$^1/_2$ pint (= 300cc).
	Front fork.	$^1/_8$ pint (= 75cc).

Exhaust system:	Low level.

Colours:	UK and general export markets -	Crystal Grey and black.
	Petrol Tank -	Crystal Grey.
	Mudguards -	Crystal Grey with black centre stripe, lined in white.
	Front Brakeplate -	Crystal Grey.
	Tank Badges -	White lettering on a black and chrome backing.
	Other painted parts -	Black.
	USA -	Aztec Red and black.
	Petrol Tank -	Aztec Red.
	Mudguards -	Aztec Red, with black centre stripe, lined in white.
	Front Brakeplate -	Aztec Red or chrome plating.
	Tank Badges -	White lettering on a black and chrome backing.
	Other painted parts -	Black.

Twinseat:	Seat top colour usually black for UK and general export and black or grey for the USA.

Notes:
New one-piece petrol tank badge from machine no. 38130. (Known as the 'Mouth Organ' badge).
New primary chain with a cast iron clutch sprocket and a new engine sprocket - all duplex items.
New lip seal fitted in the high gear bearing housing to replace the earlier felt seal and disc.
Also a new gearbox sprocket with a journal ground slightly larger than before to fit the new oil seal.
New high gear bush extending through the clutch seal. New 'labyrinth' clutch seal and new seal holder.
A cutaway in the crankcase allows a larger gearbox sprocket to be used, from no. 42865.
New rear mudguard with much deeper valance.
Steering lock housing on the frame from machine no. 36350. The lock was optional.
Rubber cover fitted to the lock housing from machine no. 37220.
New larger and more effective silencer.

The T20J was offered from 1958. It was identical to the T20 except for a restricted carburettor confining the power output to 4.95bhp @ 5,700rpm. This 'Junior' model was for sale in the USA where certain states permitted riders from the age of fourteen.

For road test figures see Appendix 8.

The production year ran from 1st August 1957 to 31st July 1958. Engine and frame numbers were always the same.

Parts List No.5 covers all 1958 (and 1957) models.

The number of production T20 & T20J Cubs built in 1958 = 7,477 T20 and 5 T20J.

Analysis of sales:	T20	UK = 67.9%.	USA = 8.9%.	Others = 23.2%.
	T20J		USA = 100%	

A 1959 T20 from the timing side. This is a 'mock up' machine fitted with 18in x 3.00in wheels. (The Mike Estall Collection)

The same 1959 T20 showing the drive side. (The Mike Estall Collection)

Appendix 1.6 Model profile

T20 Tiger Cub - 1959

T20J Junior Cub - 1959

Machines T20 or T20J 45312 (August 1958) - T20 or T20J 56359 (July 1959).

Bore & Stroke:	63mm (2.480in) x 64mm (2.520in).
Compression ratio:	7:1.
Camshaft:	Standard. Valve clearances - both 0.010in, (cold).
Ignition type:	Lucas AC/DC with battery. RM 13 alternator.
Ignition timing:	4° BTDC, fully retarded. (= 28° fully advanced).
Points gap:	0.014in - 0.016in.
Spark plug:	Champion L7, Lodge HN, KLG F80. Gap 0.025in.
Carburettor:	Zenith 17MXZ-CS5. Restrictor on T20J up to no. 47347, then Zenith 9.5mm MXZ.

Gearbox ratios:

		1st	3rd	2nd	4th
Standard ratios.					
Mainshaft		16T	25T	20T	28T
Layshaft -		29T	20T	25T	17T

Wheel & Tyre sizes:	Front.	3.25 x 16 on WM2 rim.
	Rear.	3.25 x 16 on WM2 rim.
	Tyres.	Dunlop Universal, block pattern.
Sprockets & Chains:	Engine 19T.	Clutch 48T. Primary chain $^3/_8$in x $^7/_{32}$in x 62 links, duplex.
	Gearbox 18T.	Rear wheel 46T. Rear chain $^1/_2$in x $^3/_{16}$in x 112 links.
Capacities:	Petrol.	3 galls (= 13.5 litres). New 'Humped' shaped tank. New one-piece badges.
	Oil.	2$^3/_4$ pints (= 1.56 litres).
	Gearbox.	$^1/_3$ pint (= 200cc).
	Primary Chain.	$^1/_3$ pint (= 200cc). (Previously 300cc).
	Front fork.	$^1/_8$ pint (= 75cc).

Exhaust system:	Low level.

Colours:

UK and general export markets -	Crystal Grey and black.
Petrol Tank -	Crystal Grey.
Mudguards -	Crystal Grey with black centre stripe, lined in white.
Oil Tank & Toolbox -	Crystal Grey.
Front Brakeplate -	Crystal Grey.
Side panels -	Crystal Grey.
Tank Badges -	White lettering on a black and chrome backing.
Other painted parts -	Black.
USA -	Aztec Red and black.
Petrol Tank -	Aztec Red.
Mudguards -	Aztec Red, with black centre stripe, lined in white.
Oil Tank & Toolbox -	Aztec Red.
Front Brakeplate -	Aztec Red or chrome plating.
Side panels -	Aztec Red.
Tank Badges -	White lettering on a black and chrome backing.
Other painted parts -	Black.

Twinseat:	Black, with a grey lower styling band and grey piping - all markets.

Notes:
There was a major styling change for the 1959 season; the partial enclosure of the rear of the machine by means of side panels. The oil tank and toolbox, which were accessible through holes in these panels, had altered mounting lugs to take them further back in the frame. The oil tank now did not have a drain plug so the right hand panel had to be removed in order to drain the tank.
The side panels meant that a shorter rear mudguard was used which was roughly two-thirds the length of the old one.
There was an inner mudguard in sheet steel, fixed at the front end of the rear mudguard down to the frame behind the gearbox.
A new shape of petrol tank came into use (the 'Humped' type), a design which continued right to the end of the T20 in 1966.
Anti-surge baffles were added to the inside of this tank from no. 50459.
There was a new cylinder barrel with more substantial finning - the 'Oval' barrel, with a new oval cylinder head.
New drive side flywheel now incorporated two keyways - for AC/DC or Energy Transfer systems.
Two lip seals were now used at the gearbox sprocket. Placed back-to-back, one kept the oil in, while the other kept the dirt out.
New fatter silencer with a larger diameter exit pipe. A mute was added to the silencer from no. 50825.

The T20J was a 'Junior' version sold mostly in the USA. It was identical to the T20 except for a restricted carburettor, or later a 9.5mm Zenith, confining the power output to 4.95bhp @ 5,700rpm.

The production year ran from 1st August 1958 to 31st July 1959. Engine and frame numbers were always the same.

Parts List No.6 covers all 1959 models.

The number of T20 & T20J Cubs built in 1959 = 8,923 T20 and 166 T20J.

Analysis of sales:	T20	UK = 69.8%.	USA = 6.8%.	Others = 23.4%
	T20J		USA = 94.0%.	Others = 10.0%

A 1960 T20 in Crystal Grey. This machine was restored by the author. (The Mike Estall Collection)

Appendix 1.7 Model profile

T20 Tiger Cub - 1960

T20J Junior Cub - 1960

Machines T20 or T20J 56360 (August 1959) - T20 or T20J 69516 (July 1960).

Bore & Stroke:	63mm (2.480in) x 64mm (2.520in).				
Compression ratio:	7:1.				
Camshaft:	Standard. Valve clearances - both 0.010in, (cold).				
Ignition type:	Lucas AC/DC with battery. RM 13 alternator.				
Ignition timing:	4° BTDC, fully retarded. (= 28° fully advanced).				
Points gap:	0.014in - 0.016in.				
Spark plug:	Champion L7, Lodge HN, KLG F80. Gap 0.025in.				
Carburettor:	Zenith 18MXZ-C17. 9.5mm Zenith MXZ on T20J.				
Gearbox ratios:	Standard ratios.	1st	3rd	2nd	4th
	Mainshaft	16T	25T	20T	28T
	Layshaft -	29T	20T	25T	17T
Wheel & Tyre sizes:	Front.	3.25 x 17 on WM2 rim.			
	Rear.	3.25 x 17 on WM2 rim.			
	Tyres.	Dunlop, ribbed front and block pattern rear.			
Sprockets & Chains:	Engine 19T.	Clutch 48T.	Primary chain $^3/_8$in x $^7/_{32}$in x 62 links, duplex.		
	Gearbox 17T.	Rear wheel 46T.	Rear chain $^1/_2$in x $^3/_{16}$in x 112 links.		
Capacities:	Petrol.	3 galls (= 13.5 litres). 'Humped' tank, one-piece badges.			
	Oil.	$2^3/_4$ pints (= 1.56 litres).			
	Gearbox.	$^1/_3$ pint (= 200cc).			
	Primary Chain.	$^1/_3$ pint (= 200cc).			
	Front fork.	$^1/_8$ pint (= 75cc).			

Exhaust system: Low level.

Colours:	UK and general export markets -	Crystal Grey and black.
	Petrol Tank -	Crystal Grey.
	Mudguards -	Crystal Grey with black centre stripe, lined in white.
	Oil Tank & Toolbox -	Crystal Grey.
	Front Brakeplate -	Crystal Grey.
	Side panels -	Crystal Grey.
	Tank Badges -	White lettering on a black and chrome backing.
	Other painted parts -	Black.
	USA -	Aztec Red and black.
	Petrol Tank -	Aztec Red.
	Mudguards -	Aztec Red, with black centre stripe, lined in white.
	Oil Tank & Toolbox -	Aztec Red.
	Front Brakeplate -	Aztec Red or chrome plating.
	Side panels -	Aztec Red.
	Tank Badges -	White lettering on a black and chrome backing.
	Other painted parts -	Black.

Twinseat: Black, with a grey lower styling band and grey piping - all markets.

Notes:
There were several significant changes for 1960:
A new cylinder head was used with a larger inlet valve and an exhaust port which had been moved round by seven degrees. The head had a greater counterbore for the head nuts and shorter head nuts were used. This, with longer barrel studs, allowed cylinder head removal without lowering the front of the engine.
From engine no. 57617 a new centre-split two-piece crankcase was used giving more rigid support to the crankshaft. There were new outer and inner drive and timing side covers. The distributor now has an internal clamp and the cap is secured by a self-tapping screw.
A new larger capacity brass-bodied oil pump was brought into use from engine no. 69157, together with a new timing side bush carrier with a slot to bleed oil to the skew gears.
The wheel size was put up to 17in and the gearbox sprocket changed accordingly.
The swinging arm was changed to the trials type which was wider at the front. There were longer rear suspension units and longer centre and prop stands to allow for the increased wheel diameter.
The choke size of the carburettor was increased from 17 to 18mm.
A new square rectifier, Lucas 47132B, replaced the round 47111A type.

The T20J was a 'Junior' version sold mostly in the USA. It was identical to the T20 except for a 9.5mm Zenith carburettor, confining the power output to 4.95bhp @ 5,700rpm.

The production year ran from 1st August 1959 to 31st July 1960. Engine and frame numbers were always the same, except for 70 machines where there may have been some crossover of numbers.

Parts List No.7 covers all 1960 (and 1961) models.

The number of T20 & T20J Cubs built in 1960 = 10,901 T20 and 38 T20J. This was the highest number of T20 machines built in any year.

Analysis of sales:	T20	UK = 70.9%.	USA = 8.3%.	Others = 20.8%
	T20J		USA = 100%.	

A 1961 T20 from the sales brochure. The cosmetic differences between the 1961 and 1962 home market T20 models were minimal. (The Mike Estall Collection)

Appendix 1.8 Model profile

T20 Tiger Cub - 1961 & 1962

T20J Junior Cub - 1961

Machines T20 or T20J 69517 (August 1960) - T20 88346 (July 1962).

Bore & Stroke:	63mm (2.480in) x 64mm (2.520in).
Compression ratio:	7:1.
Camshaft:	Standard. Valve clearances - both 0.010in, (cold).
Ignition type:	Lucas AC/DC with battery. RM 13 alternator. RM 18 at the very end.
Ignition timing:	4° BTDC, fully retarded. (= 28° fully advanced).
Points gap:	0.014in - 0.016in.
Spark plug:	Champion L7, Lodge HN, KLG F80. Gap 0.025in.
Carburettor:	Zenith 18MXZ-C17. Later the Amal 32/1. 9.5mm Zenith on T20J.

Gearbox ratios:

		1st	3rd	2nd	4th
Standard ratios.					
Mainshaft		16T	25T	20T	28T
Layshaft		29T	20T	25T	17T

Wheel & Tyre sizes:	Both.	3.25 x 17 on WM2 rim.
	Tyres.	Dunlop, ribbed front and block pattern rear.
Sprockets & Chains:	Engine 19T.	Clutch 48T. Primary chain ³/₈in x ⁷/₃₂in x 62 links, duplex.
	Gearbox 17T.	Rear wheel 46T. Rear chain ¹/₂in x ³/₁₆in x 112 links.
Capacities:	Petrol.	3 galls (= 13.5 litres). 'Humped' tank, one-piece badges.
	Oil.	2³/₄ pints (= 1.56 litres).
	Gearbox.	¹/₃ pint (= 200cc).
	Primary Chain.	¹/₃ pint (= 200cc).
	Front fork.	¹/₈ pint (= 75cc).

Exhaust system: Low level.

Colours, (1961):

New colour for 1961, all markets -	Silver Sheen and black.
Petrol Tank -	Black top half, Silver Sheen lower half. A chrome styling strip covers the tank seam.
Mudguards -	Silver Sheen with black centre stripe, lined in white.
Oil Tank & Toolbox -	Silver Sheen.
Front Brakeplate -	Silver Sheen, or chrome plating for the USA.
Side panels -	Silver Sheen
Tank Badges -	White lettering on a black and chrome backing. The tank badges are cut away to make room for the styling strip.
Other painted parts -	Black.

Colours (1962):

UK same as 1961 except for gold lining on the mudguards and gold lettering in the tank badges.	
New colours for 1962, all export markets -	Flame, Silver Sheen and black.
Petrol Tank -	Flame top half, Silver Sheen lower half. A chrome styling strip covers the tank seam.
Mudguards -	Silver Sheen with Flame centre stripe, lined in gold.
Oil Tank & Toolbox -	Silver Sheen.
Front Brakeplate -	Silver Sheen, or chrome plating for the USA.
Side panels -	Silver Sheen
Tank Badges -	Black lettering on a gold and chrome backing. The tank badges are cut away to make room for the styling strip.
Other painted parts -	Black.

Twinseat: Black or grey top, with a grey lower styling band and grey piping.

Notes:
Mechanical changes were that from no. 81890 a new cast iron bodied oil pump came into use, giving a better flow rate when hot. The pushrod tube internal guide plate was left out.
From engine no. 84269 a new crankcase was used, with a ball bearing supporting the timing side of the crankshaft as well as the drive side. This new crankcase still had the old distributor - this was not changed until the 1963 season.
New two-piece crankpin with a greater journal diameter of 1⁵/₁₆in. This also meant a new conrod and big end bush were used.
From engine no. 83727 an RM18 higher output alternator was used with a new rectifier, Lucas 47132 (square in shape).
The rear light unit was changed to a Lucas L564 type. No separate rear reflector now.
A new, plunger type petrol tap was used for 1962.
The Amal 32/1 carburettor was introduced from no. 81890 in the USA and at machine no. 83192 elsewhere.

The T20J 'Junior' Cub was available in the USA It was identical to the T20 with the exception of a 9.5mm Zenith carburettor which kept the power down to 4.95bhp @ 5,700rpm. This model continued to appear in the USA sales catalogue up to the end of 1965, but none were sold after 1961.

For road test figues see Appendix 8.

Each production year ran from 1st August to 31st July. Engine and frame numbers were always the same.

Parts List No.7 covers all 1961 models. No.8 should be used for 1962 (and 1963), machines.

The number of T20 Cubs built in 1961 = 9,572.
The number of T20 Cubs built in 1962 = 4,841.
There were 14 T20J sold in 1961. Twelve went to the USA and two were sent to Bermuda.

Analysis of sales:	T20 (1961)	UK = 66.6%.	USA = 3.0%.	Others = 30.4%.
	T20 (1962)	UK = 48.0%	USA = 7.8%	Others = 44.2%
	T20J (1961 only)		USA = 85.7%	Others = 14.3%.

A 1963 T20 from the left side. The UK colours were Flame and Silver Sheen, with a two level seat. (The Mike Estall Collection)

A 1965 T20 from the timing side. The colours were Hi Fi Scarlet and Silver Sheen. This machine has been restored by the author. (The Mike Estall Collection)

Appendix 1.9 Model profile

T20 Tiger Cub – 1963 onwards

Machines T20 88347 (August 1962) - T20 4011 (November 1965).

Bore & Stroke:	63mm (2.480in) x 64mm (2.520in).
Compression ratio:	7:1.
Camshaft:	Standard. Valve clearances - both 0.010in, (cold).
Ignition type:	Lucas AC/DC with battery. RM 18 alternator.
Ignition timing:	8° BTDC, fully retarded. (= 32° fully advanced).
Points gap:	0.014in - 0.016in.
Spark plug:	Champion L7, Lodge HN, KLG F80. Gap 0.025in.
Carburettor:	Amal 32/1, $^{11}/_{16}$in choke size.

Gearbox ratios:	Standard ratios.	1st	3rd	2nd	4th
	Mainshaft	16T	25T	20T	28T
	Layshaft -	29T	20T	25T	17T

Wheel & Tyre sizes:	Both.	3.25 x 17 on WM2 rim.
	Tyres.	Dunlop, ribbed front and block pattern rear.

Sprockets & Chains:	Engine 19T.	Clutch 48T.	Primary chain $^3/_8$in x $^7/_{32}$in x 62 links, duplex.
	Gearbox 17T.	Rear wheel 46T.	Rear chain $^1/_2$in x $^3/_{16}$in x 112 links.

Capacities:	Petrol.	3 galls (= 13.5 litres). 'Humped' tank, one-piece badges.
	Oil.	$2^3/_4$ pints (= 1.56 litres).
	Gearbox.	$^1/_3$ pint (= 200cc).
	Primary Chain.	$^1/_3$ pint (= 200cc).
	Front fork.	$^1/_8$ pint (= 75cc).

Exhaust system:	Low level.	

Colours:	UK and gen.export -	Flame (1963) or Hi Fi Scarlet (1964 on), Silver Sheen and black.
	Petrol Tank -	Flame (1963) Scarlet (1964 on), top half.
		Silver Sheen, lower half.
		A chrome styling strip covers the tank seam.
	Mudguards -	Silver Sheen with centre stripe in Flame (1963) or Scarlet (1964), lined in gold.
		Both mudguards ribbed 1963, front only for 1964 on.
	Oil Tank & Toolbox -	Flame (1963) or Scarlet (1964 on).
	Front Brakeplate -	Silver Sheen.
	Side panels -	Silver Sheen.
	Tank Badges -	Gold lettering on a black and chrome backing. Tank badges cut away to make room for the styling strip.
	Other painted parts -	Black.
	USA (all years) -	Flamboyant Ruby Red (later called Flamboyant Scarlet), Silver Sheen and black.
	Petrol Tank -	Red or Scarlet, top half.
		Silver Sheen, lower half.
		With chrome styling strip over the tank seam.
	Mudguards -	Silver Sheen with Red/Scarlet centre stripe, lined in gold.
		Both mudguards ribbed 1963, front only for 1964 on.
	Oil Tank & Toolbox -	Red or Scarlet.
	Front Brakeplate -	Chrome plate.
	Side panels -	Silver Sheen.
	Tank Badges -	Gold lettering on a black and chrome backing. Tank badges cut away to make room for the styling strip.
	Other painted parts -	Black.

Twinseat:	Two-level seat, grey top, with a grey lower styling band and grey piping.

Notes:

August 1962 marked the advent of the 'Side Points' engine. The points were now mounted on a backing plate located at the right hand side of the engine. The contact breaker was driven, via an automatic advance and retard unit, directly from the end of the camshaft.
There were new crankcases, reshaped internally to improve oil drainage, and new TS inner and DS outer covers.
The rocker boxes were closed with finned alloy covers.
A rubber or plastic plug was added to the timing side outer case to give easy access to the clutch cable end.
There was a new exhaust pipe with a mounting boss a few inches below the top end for a bracket to fix the pipe to the front engine mounting.
A drain plug was fitted in the oil tank so it was no longer necessary to remove the side panel when changing the oil.
There was a new rectifier - Lucas 49072 - 1^1/$_2$in diameter (black painted).
The nacelle now had separate Lucas 88SA ignition and lighting switches and the gear indicator was removed to the crankcase top, above the gearbox, for 1963 and 1964, but went out altogether for 1965.
There was a new grey faced speedometer and a new headlamp unit. At machine no. 92241 a new handlebar bend with bolt-on levers came into use. Adjustable rear suspension units were an option for 1963 only.

Each production year ran from 1st August to 31st July. Engine and frame numbers were always the same.

Parts List No.8 (Supplement) covers all 1963 models. No.9 is for 1964 models and No.10 for machines after no. 101.

The number of T20 Cubs built in 1963 - no. 88347 to 94599 = 4,696
The number of T20 Cubs built in 1964 - no. 94600 to 99719 = 3,531
The number of T20 Cubs built in 1965 on - 101 & 2001 on   = 2,721   Total = 10,948.

Analysis of sales:	1963	UK = 46.3%.	USA = 5.9%.	Others = 47.8%.
	1964	UK = 45.7%.	USA = 5.3%.	Others = 49.0%.
	1965 on	UK = 58.4%.	USA = 5.0%.	Others = 36.6%.

A 1957 T20C 'Competition Cub' from the drive side. The machine in this photo has the T20 handlebar and levers. (The Mike Estall Collection)

The T20C was the first of a long and distinguished line of sports models. This 1958 machine was restored by the author. (The Mike Estall Collection)

Appendix 1.10 Model profile

T20C Competition Cub - 1957 to 1959

Machines T20C 27660 (Jan 1957) - T20C 52974 (May 1959).

Bore & Stroke:	63mm (2.480in) x 64mm (2.520in).
Compression ratio:	7:1.
Camshaft:	Standard. Valve clearances - both 0.010in, (cold).
Ignition type:	Lucas AC/DC with battery. RM 13 alternator.
Ignition timing:	4° BTDC, fully retarded. (= 28° fully advanced).
Points gap:	0.014in - 0.016in.
Spark plug:	Champion L7 or L10S, Lodge HN or H14, KLG F70 or F80. Gap 0.025in.
Carburettor:	Amal 332/3, ³/₄in choke size, or Zenith 17MXZ-CS5 from machine no. 39167.

Gearbox ratios:

	1st	3rd	2nd	4th
Standard ratios.				
Mainshaft	16T	25T	20T	28T
Layshaft	29T	20T	25T	17T

Wheel & Tyre sizes:	Front. 3.00 x 19 on WM1 rim.
	Rear. 3.50 x 18 on WM2 rim.
	Tyres. Optional Dunlop road or trials tyres. Butted spokes on both wheels.
Sprockets & Chains:	Engine 18T. Clutch 36T. Primary chain ¹/₂in x ³/₁₆in x 48 links, simplex.
	Engine 19T. Clutch 48T. Primary chain ³/₈in x ⁷/₃₂in x 62 links, duplex. From engine no. 35847.
	Gearbox 16T. Rear wheel 54T. Rear chain ¹/₂in x ³/₁₆in x 116 links.
	Gearbox 16T. Rear wheel 46T. Rear chain ¹/₂in x ³/₁₆in x 112 links. From engine no. 35847.
Capacities:	Petrol. 2⁵/₈ gallons, (= 11.92 litres).
	Oil. 2³/₄ pints (= 1.56 litres). Froth tower on tank.
	Gearbox. ¹/₃ pint (= 200cc).
	Primary Chain. ¹/₂ pint (= 300cc).
	¹/₃ pint (= 200cc). From engine no. 45312.
	Front fork. ¹/₈ pint (= 75cc).
Exhaust system:	High level, routed outside the rear sub-frame. Downswept pipe optional. Early plunger Cub silencer used, with repositioned bracket.
Colours:	UK and gen. export - 1957 to 1959 - Crystal Grey and black.
	USA 1957 - Crystal Grey and black,
	1958 & 1959 - Aztec Red and black.
	Petrol Tank - Crystal Grey, or Aztec Red.
	Mudguards - Crystal Grey or Aztec Red, black centre stripe, lined white.
	Front - Painted alloy, plain section. Rear - Steel with centre rib.
	Oil Tank & Toolbox - Black.
	Front Brakeplate - Crystal Grey (UK), or chrome plated (USA).
	Tank Badges - White lettering on black and chrome backing.
	Other painted parts - Black.
Twinseat:	Single level seat with a black top cover and black sides, with grey lower styling band.

Notes:
Mechanically this model was identical to the T20 for 1957 to 1959, except for lower overall gearing, and incorporated the same improvements as on that model.
The T20C was the first of the sports Cubs and was based on the 1956 ISDT machine ridden by Ken Heanes. It was intended as a road machine that could easily be used off-road as well. It had road or trials tyres or gearing, high or low level exhaust pipe.
The T15 petrol tank was used but the three gallon T20 tank was also an option.
The front fork stanchions were 1in longer than those on the T20 and with longer top tapers to fit in reinforced top and bottom yokes. There was a strong front fork bridge held by two clips on each leg and a one-piece front mudguard stay in the form of a loop covering front, rear and both sides of the mudguard.
The full instrument nacelle and gear indicator were kept but the lower nacelle legs were omitted for 1958/9 when fork gaiters were used.
The machine was fitted with a sump shield.
The UK handlebars were the same as those for the USA T20 which had a higher lift than the UK T20 handlebar. There were bolt-on ball ended levers, a prop stand but no centre stand, and a new chainguard similar to the T15 type but with a separate front portion.

Production of T20C -	1957 - between no. 26276 to no. 35846 =	1,773
	1958 - between no. 35847 to no. 45311=	1,872
	1959 - between no. 45312 to no. 56359 =	466
	Total built =	4,111

Parts List No.5 covers all 1957/8 models and No.6 covers the 1959 machines.

Analysis of sales -	1957	UK = 32.1%.	USA = 60.0%.	Others = 7.9%.
	1958	UK = 37.2%.	USA = 55.2%.	Others = 7.6%.
	1959	UK = 64.8%.	USA = 19.7%.	Others = 15.5%.

An August 1957 artist's impression of the T20CA in Crystal Grey. This model did not appear until February 1958, by which time the colour had become Aztec Red. It is thought that the 3 gallon tank and one-piece badge were not used in production machines. (Courtesy John Nelson)

Appendix 1.11 Model profile

T20CA - 1958 & 1959

Machines T20CA 39773 (February 1958) - T20CA 53780 (June 1959).

Bore & Stroke:	63mm (2.480in) x 64mm (2.520in).
Compression ratio:	9:1.
Camshaft:	Standard. Valve clearances: Both 0.010in, (cold).
Ignition type:	Lucas AC/DC with battery. RM 13 alternator.
Ignition timing:	4° BTDC, fully retarded. (= 28° fully advanced). (Figures NOT confirmed).
Points gap:	0.014in - 0.016in.
Spark plug:	Champion L7 or L10S, Lodge HN or H14, KLG F70 or F80. Gap 0.025in.
Carburettor:	Zenith 17MXZ-CS5.

Gearbox ratios:	Close ratios.	1st	3rd	2nd	4th
	Mainshaft	16T	25T	20T	27T
	Layshaft -	29T	20T	25T	18T

Wheel & Tyre sizes:	Front.	3.00 x 19 on WM1 rim.
	Rear.	3.50 x 18 on WM2 rim.
	Tyres.	Dunlop Trials Universal, then road tyres from no. 45312.
Sprockets & Chains:	Engine 19T.	Clutch 48T. Primary chain $^3/_8$in x $^7/_{32}$in x 62 links, duplex.
	Gearbox 16T.	Rear wheel 46T. Rear chain $^1/_2$in x $^3/_{16}$in x 112 links.
Capacities:	Petrol.	2$^5/_8$ gallons, (= 11.92 litres). T15 'Teardrop' shape tank.
	Oil.	2$^3/_4$ pints (= 1.56 litres). Froth tower on tank.
	Gearbox.	$^1/_3$ pint (= 200cc).
	Primary Chain.	$^1/_2$ pint (= 300cc).
		$^1/_3$ pint (= 200cc) from engine no. 45312.
	Front fork.	$^1/_8$ pint (= 75cc).

Exhaust system:	Low level.

Colours:	Aztec Red and Black.	
	Petrol Tank -	Aztec Red.
	Mudguards -	Aztec Red, with black centre stripe lined in white.
		Front - Painted alloy, plain section.
		Rear - Steel, with centre rib.
	Oil Tank & Toolbox -	Black.
	Front Brakeplate -	Chrome plate.
	Tank Badges -	White lettering on black and chrome backing.
	Other painted parts -	Black.

Twinseat:	Black top cover with black sides and grey lower styling band.

Notes:
The T20CA was similar to the T20C but had a low level exhaust system and trials or road tyres, and was intended as an on/off road sports machine. The engine was fitted with a high compression piston and close ratio gears.
Mechanically, except for the compression ratio and internal gearing, the machine was identical to the T20 and the T20C and incorporated all the improvements given to those models:
1958 - Duplex primary chain and a cast iron clutch drum, a garter seal on the gearbox sprocket and a longer high gear bush extending through the clutch oil seal. The steering lock was available on all T20CA and a Zenith carburettor was used from the start of the model.
1959 - New gearbox sprocket oil seal arrangement - two seals placed back to back. The new 'Oval' cylinder barrel with increased finning area and a new cyclinder head secured by shorter nuts.
It had front forks and nacelle taken from the T20C with stanchions 1in longer than those on the T20 and with longer top tapers to fit in reinforced top and bottom yokes. There was a strong fork front bridge held by two clips on each leg and a one-piece front mudguard stay in the form of a loop covering front, rear and both sides of the mudguard.
The mudguards were sports 'Trophy' style items, the front being painted alloy and plain in section whereas the rear was ribbed in the centre.
The instrument nacelle was kept but the lower legs were omitted and replaced with rubber gaiters.
The machine was fitted with a crankcase undershield, high lift handlebars with bolt-on levers, and no centre stand.

A large proportion of all T20CA machines were sold in the USA but a few went to Canada, Ceylon, Iran, Puerto Rico, Mexico, Tanganyika and Finland.

1958 - no. 35847 to no. 45311 - number of T20CA built =	14
1959 - no. 45312 to no. 56359 - number of T20CA built =	351
The total number of T20CA built =	365

Parts List No.5 covers all 1958 models and No.6 covers 1959; however, the T20CA was a minor model not specifically mentioned.

Analysis of sales -	1958	USA = 42.9%	Others = 57.1%.
	1959	USA = 96.6%	Others = 3.4%.

The T20S road variant from the timing side. This example has the Zenith carburettor instead of the more usual Monobloc. (The Mike Estall Collection)

The same T20S roadster from the drive side. (The Mike Estall Collection)

Appendix 1.12 Model profile

T20S Scrambler - 1959 to 1961

T20W Woods - 1960

Machines T20S 48139 (December 1958) - T20S 75973 (March 1961) & T20W 60265 - 60290 (January 1960).

Bore & Stroke:	63mm (2.480in) x 64mm (2.520in).
Compression ratio:	9:1 (7:1 on trials engine, early home market engines and on the T20W).
Camshaft:	Sports - 'R' type. Valve clearances - Inlet 0.002in, Exhaust 0.004in, (cold).
Ignition type:	Lucas Energy Transfer. RM 13 Alternator.
Ignition timing:	7:1 c.r. - 8° BTDC, (= 32° fully advanced). Rotor on keyway no. 2, (7 o'clock).
	9:1 c.r. - 16° BTDC, (= 40° fully advanced). Rotor on keyway no. 1, (3 o'clock).
Points gap:	0.014in - 0.016in.
Spark plug:	Champion L7, L5 or LA11, Lodge HN, HN3 or H14, KLG F70 or F100. Gap 0.020in.
Carburettor:	Limited information available, see Notes below.
	Amal 332 $^7/_8$in choke size or Monobloc 376/217 $^{15}/_{16}$in choke size with remote float chamber or Monobloc 376/272 $^{15}/_{16}$in choke size.
	Also 18 or 20mm Zenith for certain uses (probably low comp. engines).

Gearbox ratios: Optional - Wide (W) or Close (C) or Extra Close Ratios (ECR).

	1st - All	3rd - W	C	ECR	2nd - All	4th - W	C	ECR
Mainshaft -	16T	25T	25T	23T	20T	29T	27T	25T
Layshaft -	29T	20T	20T	22T	25T	16T	18T	20T

Wheel & Tyre sizes:	Front.	3.00 x 19 on WM1 rim.
	Rear.	3.50 x 18 on WM2 rim.
	Tyres.	Dunlop tyres as appropriate to the type of machine.
Sprockets & Chains:	Engine 19T.	Clutch 48T. Primary chain $^3/_8$in x $^7/_{32}$in x 62 links, duplex.
	Gearbox 16 - 19T.	Rear wheel 46 - 58T. Rear chain $^1/_2$in x $^3/_{16}$in x 112 - 117 links.
Capacities:	Petrol.	2$^5/_8$ gallons, (= 11.92 litres). T15 'Teardrop' shape tank.
	Oil.	2$^3/_4$ pints (= 1.56 litres). Froth tower on tank.
	Gearbox.	$^1/_3$ pint (= 200cc).
	Primary Chain.	$^1/_3$ pint (= 200cc).
	Front fork.	$^1/_4$ pint (= 150cc).
Exhaust system:	High level, outside the rear sub-frame. Two types - for use with or without tachometer.	

Colours:

	UK and general export.	USA
1959 -	Ivory and Azure Blue.	Ivory with Azure Blue or Aztec Red.
1960 -	Ivory and Azure Blue.	Ivory and Azure Blue.
1961 -	Ruby Red and Silver Sheen.	*Ruby Red and Silver Sheen. *(Known as Bright Amaranth Red in the USA.)
Petrol Tank -	top - first colour,	
	bottom - second colour. n.b. - the colours were sometimes reversed in 1959.	
Mudguards -	1959/60 - first colour, with second colour centre stipe, front lined in gold.	
	1961 - second colour with first colour centre stripe, front lined in gold.	
	Front - Painted alloy, plain section. Rear - Steel, with centre rib, no gold lining.	
Oil Tank & Toolbox -	Black.	
Front Brakeplate -	Silver Sheen, (UK). Black or Silver Sheen (USA).	
Tank Badges -	White lettering on black and chrome backing.	
Other painted parts -	Black.	

Twinseat: Single level seat. Grey or black top cover with black sides and grey lower styling band.

Notes:
The T20S was a developement of the T20C Competition Cub and was available in three different guises:
1. Standard/Road spec. - Lights, 'D' shaped speedometer or a round speedometer with tachometer, road tyres, 9:1 piston, 'R' cam, CR gears, silencer, small toolbox, $^{15}/_{16}$in Monobloc 376/272 or Zenith carburettor.
2. Trials spec. - No lights, optional speedometer, trials tyres, 7:1 piston, 'R' cam, WR gears, silencer and a $^7/_8$in Amal 332, Monobloc or Zenith. Some early machines may have had standard gears - WR set was perhaps not available until May 1959.
3. Scrambles/Racing spec. - No lights, horn, number plates or speedometer, scrambles tyres, 9:1 piston, 'R' cam, CR or ECR gears, exhaust pipe extension and a 140 or 200 main jet in the Monobloc 376/217 with a remote float chamber.
All models had a high level exhaust system (outside the sub-frame), prop stand only, a crankcase undershield, heavyweight forks (internally sprung), ET ignition systems, sports valve springs, 'R' cam, and the large inlet valve.

The T20W was intended for reliability trials, cross country events and the like. It had a high level exhaust, WR gears, low compression piston, standard cam, 18mm Zenith carburettor and Dunlop Trials Universal tyres. The factory despatch books show only six sold but the Tiger Cub and Terrier Register has details of others, so the actual number built is not known. The model was replaced by the T20T.

Production of T20S:	1959 - no. 45312 to 56359 =	1,074	
	1960 - no. 56360 to 69516 =	2,098	(plus at least 6 T20W)
	1961 - no. 69517 to 81889 =	320	
	The total number of T20S built =	3,492	(plus 1 in 1968) = 3493

The first T20S was no. 48139, a low compression ratio model despatched to a UK customer. It was the first of a batch of 197 machines built in December 1958, most of which went to the USA. The last was no. T20S 8137, which was sent to Mozambique in March 1968.

Parts List No.6 covers all 1959 models and No.7 covers 1960 and 1961.

Analysis of Sales:	1959	UK = 14.3%.	USA = 79.5%.	Others = 6.2%.
	1960	UK = 34.4%.	USA = 59.3%.	Others = 6.3%.
	1961		USA = 95.0%.	Others = 5.0%.

The T20T fully kitted-out with lights. This artist's impression is shown with Zenith carburettor. (The Mike Estall Collection)

Appendix 1.13 Model profile

T20T Trials - 1961

Machines T20T 56145 (September 1959) - T20T 833 (November 1964).

Bore & Stroke:	63mm (2.480in) x 64mm (2.520in).
Compression ratio:	7:1.
Camshaft:	Standard. Valve clearances - both 0.010in, (cold).
Ignition type:	Lucas Energy Transfer. RM 13 alternator.
Ignition timing:	8º BTDC, (= 32º fully advanced). Rotor on keyway no. 2, (7 o'clock).
Points gap:	0.014in - 0.016in.
Spark plug:	Champion L7, Lodge HN, KLG F80. Gap 0.020in.
Carburettor:	Zenith 18MXZC18 on JoMo and home market machines or Monobloc 375/44, $^{25}/_{32}$in choke size, on TriCor models. Vokes 'D' shaped air filter.

Gearbox ratios:

	1st	3rd	2nd	4th
Wide ratios:				
Mainshaft	16T	25T	20T	29T
Layshaft	29T	20T	25T	16T

Wheel & Tyre sizes:	Front.	3.00 x 19 on WM1 rim.
	Rear.	3.50 x 18 on WM2 rim.
	Tyres.	Dunlop Universal trials.
Sprockets & Chains:	Engine 19T.	Clutch 48T. Primary chain $^3/_8$in x $^7/_{32}$in x 62 links, duplex.
	Gearbox 16T.	Rear wheel 54T. Rear chain $^1/_2$in x $^3/_{16}$in x 116 links.
Capacities:	Petrol.	$2^5/_8$ gallons, (= 11.92 litres). T15 'Teardrop' shape tank.
	Oil.	$2^3/_4$ pints (= 1.56 litres). Froth tower on tank.
	Gearbox.	$^1/_3$ pint (= 200cc).
	Primary Chain.	$^1/_3$ pint (= 200cc).
	Front fork.	$^1/_4$ pint (= 150cc).

Exhaust system: High level, routed outside the rear sub-frame, with silencer. Two types of pipe - for use with or without the optional tachometer.

Colours:	Ruby Red* & Silver Sheen. (*Also known as Bright Amaranth Red in the USA).
Petrol Tank -	top half - Ruby Red.
	bottom half - Silver Sheen.
Mudguards -	Silver Sheen, with Ruby Red centre stripe, front lined in gold.
	Front - Painted alloy, plain section. Rear - Steel, with centre rib, no gold lining.
Oil Tank & Toolbox -	Black.
Front Brakeplate -	Silver Sheen.
Tank Badges -	White lettering on black and chrome backing.
Other painted parts -	Black.

Twinseat: Single level seat with a grey or black top cover, black sides and grey lower styling band.

Notes:
The T20T was intended as an off-road/Enduro machine and was known as the 'Trials & Woods' model in the USA. It was announced in October 1960 along with the T20S and the T20SL, but the first deliveries of the T20T did not take place until December 1960. It was built for the 1961 year only (apart from a very few machines despatched in 1959, 1960, 1962 & 1965). There was also some UK military involvement as 40 machines, suitably modified for army use, were sold to the War Office in July 1961. These were painted a dark gloss green, known as Deep Bronze Green. They had modifications which were later incorporated into the T20M.WD French Army Model.
Apart from wide ratio gears, Energy Transfer ignition, a smaller gearbox sprocket and a cylinder head with a larger inlet valve, the engine unit was the same as for the standard T20. There were optional lights with a quickly detachable headlamp unit, a prop stand only, a crankcase undershield and a small triangular shaped toolbox. The handlebars were upswept and fitted with ball-ended levers.
Some evidence suggests that a tachometer option was available. If that was the case there would have been a different timing side outer cover incorporating a tachometer drive and two matching black-faced, circular instruments. Otherwise there was a 'D' shaped chronometric speedo only. The exhaust pipe (which always went outside the rear sub-frame), would have been a different shape depending on whether or not a tachometer drive timing side outer cover was used.

T20T deliveries:		
	1959 - no. 45312 to no. 56359 =	1
	1960 - no. 56360 to no. 69516 =	2
	1961 - no. 69517 to no. 81889 =	620
	1962 - no. 81890 to no. 88346 =	1
	1965 - no. 99733 to no. 100013 and no. 101 on =	7
	The total number of T20T built =	631

The first T20T left the factory in December 1960 and the last in November 1964. The five of these last seven machines were delivered to Triumph's main agent in Pakistan; the other two were delivered within the UK.

Parts List No.7 covers all 1961 T20T models.

Analysis of sales:	1959 -	UK = 100%.		
	1960 -	UK = 100%.		
	1961 -	UK = 44.8%.	USA = 47.4%.	Others = 7.8%.
	1962 -	UK = 100%.		
	1965 -	UK = 28.6%		Others = 71.4%.

The T20SL in road trim. (The Mike Estall Collection)

Appendix 1.14 Model profile

T20SL Scrambler Lights - 1961

Machines T20SL 64592 (April 1960) - T20SL 81889 (July 1961).

Bore & Stroke:	63mm (2.480in) x 64mm (2.520in).
Compression ratio:	9:1.
Camshaft:	Sports - 'R' type. Valve clearances - Inlet 0.002in, Exhaust 0.004in (cold).
Ignition type:	Lucas Energy Transfer. RM 13 alternator.
Ignition timing:	16º BTDC (= 40º fully advanced). Rotor on keyway no. 1 (3 o'clock).
Points gap:	0.014in - 0.016in.
Spark plug:	Champion L7 or L5, Lodge HN or H14, KLG F70 or F80. Gap 0.020in.
Carburettor:	Monobloc ¹⁵/₁₆in choke 376/272 - UK and USA east coast. Monobloc ¹⁵/₁₆in choke 376/317 - USA west coast.

Gearbox ratios:	Close ratios.	1st	3rd	2nd	4th
	Mainshaft	16T	25T	20T	27T
	Layshaft	29T	20T	25T	18T

Wheel & Tyre sizes:	Front.	3.00 x 19 on WM1 rim.
	Rear.	3.50 x 18 on WM2 rim.
	Tyres.	Dunlop road or 'knobblies'.

Sprockets & Chains:	Engine 19T.	Clutch 48T.	Primary chain ³/₈in x ⁷/₃₂in x 62 links, duplex.
	Gearbox 17T.	Rear wheel 48T.	Rear chain ¹/₂in x ³/₁₆in x 113 links.

Capacities:	Petrol.	2⁵/₈ gallons, (= 11.92 litres). T15 'Teardrop' shape tank.
	Oil.	2³/₄ pints (= 1.56 litres). Froth tower on tank.
	Gearbox.	¹/₃ pint (= 200cc).
	Primary Chain.	¹/₃ pint (= 200cc).
	Front fork.	¹/₄ pint (= 150cc).

Exhaust system:	Low level or, in scrambler version, high level and routed outside the rear sub-frame. Two types of high level pipe, for use with or without the optional tachometer.

Colours:	Ruby Red* & Silver Sheen. (*Also known as Bright Amaranth Red in the USA).
	Petrol Tank - top half - Ruby Red. lower half - Silver Sheen.
	Mudguards - Silver Sheen, with Ruby Red centre stripe, lined in gold. Front - Painted alloy, plain section. Rear - Steel, with centre rib.
	Oil Tank & Toolbox - Black.
	Front Brakeplate - Silver Sheen.
	Tank Badges - White lettering on black and chrome backing.
	Other painted parts - Black.

Twinseat:	Single level seat with a grey or black top cover, black sides and grey lower styling band.

Notes:
The T20SL was a direct developement of the American market scrambler version of the T20S. Lights were added, hence the model type T20SL. There were two versions:
1. Road version - Downswept exhaust pipe and silencer, 'Bonneville' style flat profile handlebar, road tyres, small triangular toolbox, either a 'D' shaped speedometer or a round speedometer with a tachometer, lights with a quickly detachable headlamp.
2. Scrambler version - Upswept exhaust pipe with either a silencer or an extension pipe, upswept handlebar, off-road tyres, no toolbox or lights.
Both versions had the 9:1 piston, 'R' cam and sports type valve springs, close ratio gears (extra close ratios may have been an option on the scrambler type), Energy Transfer ignition system, Monobloc carburettor, prop stand only, large inlet valve, internally sprung heavyweight front forks, ball ended handlebar levers and a crankcase shield.

For road test figures see Appendix 8.

The T20SL was only made for the 1961 season. The first machine was no. 64592. This was in fact a number from the previous, (1960), season but the machine did not leave the factory until November 1960, well into the 1961 'year'. The last T20SL was no. 81889, (which was also the last number for 1961), and this machine was sent to Eire in September 1961.

Parts List No.7 covers the T20SL.

A total of 1,814 T20SL were built.

Analysis of sales: 1961   UK = 70.9%.   USA = 23.7%.   Others = 5.4%.

An artist's impression of a side-points USA west coast T20SS with the optional high level exhaust. This model was originally the T20SL renamed and went on to become the east coast T20SC. (The Mike Estall Collection)

Appendix 1.15 Model profile

T20SS Street Scrambler - 1962 to 1965

Machines T20SS 82276 (September 1961) - T20SS 1715 (January 1965).

Bore & Stroke:	63mm (2.480in) x 64mm (2.520in).
Compression ratio:	9:1. (7:1 option available).
Camshaft:	Sports - 'R' type. Valve clearances - Inlet 0.002in, Exhaust 0.004in, (cold).
Ignition type:	Lucas Energy Transfer. RM 13 alternator, from engine no. 83727 an RM 19.
Ignition timing:	20º BTDC, (= 40º fully advanced). Locate rotor on keyway no. 1 (3 o'clock).
Points gap:	0.014in - 0.016in.
Spark plug:	Champion L7. Gap 0.020in.
Carburettor:	Monobloc 376/272, $^{15}/_{16}$in choke size, with pancake air filter.
Gearbox ratios:	Optional - Close (CR) or Extra Close Ratios (ECR).

	1st - All	3rd - CR	ECR	2nd - All	4th - CR	ECR
Mainshaft	16T	25T	23T	20T	27T	25T
Layshaft	29T	20T	22T	25T	18T	20T

Wheel & Tyre sizes:	Front.	3.00 x 19 on WM1 rim.
	Rear.	1962/3 - 3.50 x 18 on WM2 rim.
		1964/5 - 4.00 x 18 on WM3 rim.
	Tyres.	Dunlop road or off-road.
Sprockets & Chains:	Engine 19T.	Clutch 48T. Primary chain $^3/_8$in x $^7/_{32}$in x 62 links, duplex.
	Gearbox 17T.	Rear wheel 48T. Rear chain $^1/_2$in x $^3/_{16}$in x 113 links.
Capacities:	Petrol.	2$^5/_8$ gallons (= 11.92 litres). T15 'Teardrop' shape tank.
	Oil.	2$^3/_4$ pints (= 1.56 litres). Froth tower on tank.
	Gearbox.	$^1/_3$ pint (= 200cc).
	Primary Chain.	$^1/_3$ pint (= 200cc).
	Front fork.	$^1/_4$ pint (= 150cc), 1962/4.
		$^1/_3$ pint (= 200cc), 1965.

Exhaust system:	Low level.
	High level option - two types, both go outside the sub-frame, with or without tachometer for 1962, or a third pattern to go inside the sub-frame from 1963.

Colours:	All markets	
	1962 -	Burgundy and Silver Sheen.
	1963 -	Flamboyant Flame and Silver Sheen.
	1964 -	Crystal Blue and Silver Sheen.
	1964 -	Kingfisher Blue and Silver Sheen.
	Petrol Tank -	top half - first colour.
		lower half - Silver Sheen.
	Mudguards -	Silver Sheen, with first colour centre stripe, lined gold.
	Front -	Painted alloy, plain section.
	Rear -	Steel, with centre rib for 1962/3 only. Optional polished mudguards in 1964/5.
	Oil Tank & Toolbox -	Black.
	Front Brakeplate -	Silver Sheen.
	Tank Badges -	Black lettering on gold and chrome backing.
	Other painted parts -	Black.

Twinseat:	1962 - single level, 1963/5 two-level. Grey top cover with black sides and grey lower styling band.

Notes:
The T20SS was a USA west coast model for JoMo but the east coast TriCor T20SC model was also built to the same specification for 1963 onwards. It was a continuation of the 1961 T20SL and was an all purpose road/sports model intended primarily for the USA. It had Energy Transfer ignition, lights, close ratio gears (extra close for some export markets), the 9:1 piston (optional 7:1), 'R' cam, sports valve springs, optional tachometer, heavyweight forks, the small Terrier type petrol tank, a large pancake air cleaner, the small triangular toolbox, crankcase undershield, prop and centre stands. It had upswept handlebars with ball-ended levers. The 1962 model had a distributor engine but machines from no. 88347 were side points type.
In 1963 and 1964 adjustable rear suspension units were fitted, and from 1963 there were new pattern exhaust pipes, high and low level. From 1964 a new front frame with a double gusset at the headstock came into use and a new rear sub-frame with a right hand side tube cranked at the top and bottom to facilitate fitting of a high level exhaust pipe inside the sub-frame. From 1964 only the high level pipe was listed although contemporary pictures show the low level system too. A new deeper section chainguard, with protection for the lower chain run, was listed for 1964 onwards. 1965 brought in a new design of front fork with the springs outside the stanchions, also a new highly cranked splined kickstart lever and the Lucas L679 rear light unit to replace the earlier L564 type.
In 1962 the instruments were black-faced with the optional tachometer driven from the end of the camshaft via a special timing side outer cover. From 1963 the tachometer drive came out of the old distributor housing to a black-faced instrument with a matching speedometer, but if there was no tachometer fitted the speedometer was grey-faced. From 1964 all the instruments were grey-faced.

Production of T20SS:	1962 - no. 81890 to no. 88346 =		576
	1963 - no. 88347 to no. 94599 =		553
	1964 - no. 94600 to no. 99732 =		56
	1965 - no. 99733 to no. 100013 and no. 101 on =		8
	The total number of T20SS built =		1,193

Analysis of sales:	1962	UK = 2.3%.	USA = 81.4%.	Others = 17.3%.
	1963	UK = 0.2%	USA = 85.0%.	Others = 14.8%.
	1964	None	USA = 44.6%.	Others = 55.4%.
	1965	UK = 37.5%	None	Others = 62.5%.

The 1962 T20SH. Note the crankcase undershield - fitted to this model in 1962 and 1963 only. This particular machine resided in the Meriden showroom before being broken up and used as service exchange spares. (The Mike Estall Collection)

An artist's impression of the home market 1965 T20SH. Note the splined kick-start lever used for 1965 only. (The Mike Estall Collection)

Appendix 1.16 Model profile

T20SH Sports Home - 1962 to 1966

Machines T20SH 82756 (October 1961) - T20SH 4373 (January 1966).

Bore & Stroke:	63mm (2.480in) x 64mm (2.520in).
Compression ratio:	9:1.
Camshafts:	Sports - 'R' type. Valve clearances - Inlet 0.002in, Exhaust 0.004in, (cold).
Ignition type:	Lucas AC/DC with battery. RM 18 alternator.
Ignition timing:	16° BTDC, (= 40° fully advanced). Rotor on keyway no. 1, (3 o'clock).
Points gap:	0.014in - 0.016in.
Spark plug:	Champion L7 or L5, Lodge HN or H14, KLG F70 or F80. Gap 0.025in.
Carburettor:	Monobloc 376/272, $^{15}/_{16}$in choke size.

Gearbox ratios:

	1st	3rd	2nd	4th
Close ratios.				
Mainshaft	16T	25T	20T	27T
Layshaft	29T	20T	25T	18T

Wheel & Tyre sizes:	Front. 3.00 x 19 on WM1 rim.
	Rear. 3.50 x 18 on WM2 rim.
	Tyres. Dunlop road.
Sprockets & Chains:	Engine 19T. Clutch 48T. Primary chain $^3/_8$in x $^7/_{32}$in x 62 links, duplex.
	Gearbox 17T. Rear wheel 48T. Rear chain $^1/_2$in x $^3/_{16}$in x 113 links.
Capacities:	Petrol. 3 galls (= 13.5 litres). 'Humped' tank, one-piece badges.
	Oil. 2$^3/_4$ pints (= 1.56 litres). Froth tower on tank.
	Gearbox. $^1/_3$ pint (= 200cc).
	Primary Chain. $^1/_3$ pint (= 200cc).
	Front fork. $^1/_4$ pint (= 150cc), 1962/4.
	$^1/_3$ pint (= 200cc), 1965/6.

Exhaust system:	1962 - Low level, without a boss for a bracket to the front engine mount. High level option.
	1963/6 - Low level only, now with a boss on the exhaust pipe for a bracket to the front engine mount.

Colours:	1962/5 -	First colour = Burgundy (1962/3) or Hi Fi Scarlet (1964/5). Second colour = Silver Sheen.
	1966 -	First colour = Metallic Blue. Second colour = Alaskan White. (Some had 1965 colours).
	Petrol Tank-	top half - First colour.
		lower half - Second colour. Chrome styling strips over the seam.
	Mudguards, 1962/5 -	Second colour, centre stripe in the first colour, lined in gold.
		Front - painted alloy. Rear - steel, ribbed in the centre for 1962/3 only.
	1966 -	First colour, centre stripe in second colour, without a gold lining.
		Both mudguards were painted steel and without a central rib.
	Oil Tank, toolbox & switch brackets -	Red/silver models - black.
		Blue/White models - Metallic Blue.
	Front Brakeplate -	Red/silver models - silver.
		Blue/white models -Blue.
	Tank Badges -	Red/silver models - white lettering on black and chrome backing.
		Blue/white models - blue lettering on white and chrome backing.
	Other painted parts -	Black.
Twinseat:	Two level, grey top cover with black sides and grey lower styling band.	

Notes:
Announced in February 1962, the T20SH was a development of the T20SS and T20SL models, and was a pure sports roadster. It was intended mainly for the home market (although a few machines went to other export markets and a mere handful were sent to the USA). It had the T20 three gallon petrol tank, a battery box and a QD front headlamp shell. In the engine a 9:1 piston ran with the 'R' cam and close ratio gears. The crankshaft turned on two ball races in all T20SH machines. The engine was fitted with all the new mechanical improvements as they arrived - for example, side points for 1963, a square head and barrel from February 1965 and the slider block driven oil pump from August 1965. The ignition system was the rectified AC/DC type with battery. The cylinder head had the large inlet valve and sports type valve springs, and a large choke Monobloc carburettor was used. There was a standard front frame up to 1964 but after that the front frame loop had double gussets under the headstock. The right side tube of the rear sub-frame was cranked outwards at the bottom only up to 1964, but from 1965 it became cranked at the top and bottom. Heavyweight front forks were fitted. A 'D' shaped speedo was fitted or, if the tachometer option was taken up, a matching round speedo was used. The tachometer was driven from the end of the camshaft in 1962 and a special timing cover used. From 1963 the tachometer drive was taken from the old distributor position. There were alternative flat or upswept handlebars with plain or ball-ended levers. The machine had both prop and centre stands, and a crankcase undershield up to 1964

For road test figures see Appendix 8.

Parts List No. 8 covers the T20SH for 1962 and 1963, No. 9 covers 1964, No. 1 for machines after no. 101 (August 1964).

Production of T20SH:		
1962 - no. 81890 to no. 88346 =		604
1963 - no. 88347 to no. 94599 =		363
1964 - no. 94600 to no. 99732 =		226
1965 on - no. 99733 to no. 100013 and no. 101 on =		633
The total number of T20SH built =		1,826

Analysis of sales:				
	1962	UK = 96.5%.	USA = 0.2%.	Others = 3.3%.
	1963	UK = 93.7%	None	Others = 6.3%.
	1964	UK = 92.9%	USA = 1.8%.	Others = 5.3%.
	1965 on	UK = 90.8%	USA = 0.1%	Others = 9.1%.

A T20SR as shown in the 1962 sales brochure. (The Mike Estall Collection)

This picture of the T20SC is also from the 1962 sales brochure. (The Mike Estall Collection)

Appendix 1.17 Model profile

T20SC & T20SR Scrambler Comp. & Road - 1962 to 1965

Machines T20SC 84912 - T20SC 2034 & T20SR 84967 - T20SR 2029 (March 1962 - July 1965).

Bore & Stroke:	63mm (2.480in) x 64mm (2.520in).
Compression ratio:	9:1 (but the USA west coast models had a 7:1 piston for 1965).
Camshaft:	Sports 'R' type. Valve clearances - Inlet 0.002in. Exhaust 0.004in, (cold).
Ignition type:	Lucas Energy Transfer. RM 19 alternator. (The 1965 west coast T20SR - RM18 AC/DC alternator and battery).
Ignition timing:	20° BTDC, (= 40° fully advanced). Rotor on keyway no. 1 (3 o'clock).
Points gap:	0.014in - 0.016in.
Spark plug:	Champion L7. Gap 0.020in for ET models, 0.025in for battery models.
Carburettor:	Monobloc 376/272. (Or 376/314 for T20SR USA west coast 1965). Both are $^{15}/_{16}$in choke size.
Gearbox ratios:	T20SR = Close or Extra Close ratios. T20SC = Standard ratios 1962, and Wide ratios 1963/5.

Standard, (S), Close, (C), Wide, (W), or Extra Close Ratios, (ECR).

	1st - All	3rd - S	C	W	ECR	2nd - All	4th - S	C	W	ECR
Mainshaft	16T	25T	25T	25T	23T	20T	28T	27T	29T	25T
Layshaft -	29T	20T	20T	20T	22T	25T	17T	18T	16T	20T

Wheel & Tyre sizes:	Front.	3.00 x 19 on WM1 rim.
	Rear.	3.50 x 18 on WM2 rim, except - T20SC - 4.00 x 18 on WM3 rim, 1964/5.
	Tyres.	Dunlop. T20SC used Trials Universal. T20SR used K70 Gold Seal.
Sprockets & Chains:	Engine 19T.	Clutch 48T. Primary chain $^3/_8$in x $^7/_{32}$in x 62 links, duplex.
T20SC	Gearbox 17T.	Rear wheel 54T. Rear chain $^1/_2$in x $^3/_{16}$in x 116 links.
T20SR	Gearbox 17T.	Rear wheel 48T. Rear chain $^1/_2$in x $^3/_{16}$in x 113 links.
Capacities:	Petrol.	$2^5/_8$ gallons (= 11.92 litres). T15 'Teardrop' shape tank.
	Oil.	$2^3/_4$ pints (= 1.56 litres). Froth tower on tank.
	Gearbox.	$^1/_3$ pint (= 200cc).
	Primary Chain.	$^1/_3$ pint (= 200cc).
	Front fork.	$^1/_4$ pint (= 150cc) 1962/4.
		$^1/_3$ pint (= 200cc) 1965.

Exhaust system:	T20SC - High level with silencer or extension pipe. T20SR - Low level with silencer.

Colours:	1962 -	Both models - First colour - Burgundy.
	1963 -	Both models - First colour - Flamboyant Flame.
	1964 -	Both models - First colour - Kingfisher Blue or Hi Fi Scarlet.
	1965 -	T20SR - First colour - Pacific Blue.
	1965 -	T20SC - Hunting Yellow and black.
	All	(Not 1965 T20SC) - Second colour - Silver Sheen.
	Petrol Tank -	top half - First colour or Hunting Yellow in 1965.
		lower half - Second colour or Hunting Yellow in 1965.
	Mudguards -	Second colour, with first colour centre stripe lined in gold.
		Front - painted alloy. Rear - steel, ribbed in the centre for 1962/3.
		Polished alloy, T20SC for 1964/5.
	Oil Tank & Toolbox - Black.	
	Front Brakeplate:	
	1962/4 -	T20SC – Black. T20SR - Silver Sheen.
	1965 -	Silver Sheen on both models.
	Tank Badges -	Black lettering on gold & chrome backing.
	Other painted parts -	Black.

Twinseat:	Two level. T20SR - grey top cover with black sides and grey lower styling band. T20SC - all black.

Notes:
These two models were for the sporting rider. The T20SR was a 'hot' roadster with all the high performance parts - high compression piston, sports cam, CR or ECR gears, sports valve springs, large inlet valve, a large bore Monobloc carburettor, low level exhaust and road tyres. The T20SC was much the same but more suited for cross-country, on/off road use. The T20SC had a high level exhaust, SR or WR gears and trials tyres.
Both models had a crankcase undershield, heavyweight front forks (externally sprung for 1965), high rise handlebars with bolt-on levers, Energy Transfer ignition with detachable lights and a speedometer. For 1965 only there was a heavily cranked, pivoting kickstart lever. The T20SR also had the option of a tachometer and neither model had a centre stand. For 1962 both models came with a distributor motor, but from 1963 the side points engine was used. All were fitted with the 'Oval' head and barrel. The rear light unit was a Lucas L564 except for the east coast USA models in 1965, which used the L679 type. Adjustable rear suspension units were listed for 1965.
These machines were sold in export markets only and almost entirely to the east coast of the USA, except for 1965 when about two-thirds of the production went to west coast dealers. A few were sold to other countries (but not the UK). The T20SC was the east coast equivalent of the west coast T20SS and it later became the basis for the T20SM Mountain Cub.

Number of machines built -		T20SC	T20SR
1962 - no. 81890 to no. 88346		= 60	= 36
1963 - no. 88347 to no. 94599		= 137	= 74
1964 - no. 94600 to no. 99732		= 153	= 133
1965 - no. 99733 to no. 100013 & no. 101 on		= 261	= 571
Totals -		= 611	= 814

Parts List No.8 and Supplement cover 1962/3 models, No.9 deals with 1964 models, No.10 is for 1965. However, the T20SR and SC have only a passing mention in any of these publications.

		T20SC		T20SR	
Analysis of sales:	1962	USA = 100%	Others = None	USA = 100%	Others = None
	1963	USA = 100%	Others = None	USA = 100%	Others = None
	1964	USA = 96.1%	Others = 3.9%	USA = 94.7%	Others = 5.3%
	1965	USA = 100%	Others = None	USA = 99.3%	Others = 0.7%

A 1962 TR20 fitted with some road equipment. A very purposeful looking machine which even today is a most desirable model. (The Mike Estall Collection)

The very rare TS20 scrambler Cub. Note the use of an extension pipe instead of a silencer. (Courtesy EMAP)

Appendix 1.18 Model profile

TR20 Trials & TS20 Scrambler - 1962 to 1965

Machines TR20 85108 - TR20 1867 and TS20 85323 - TS20 1870 (March 1962 - February 1965).

Bore & Stroke:	63mm (2.480in) x 64mm (2.520in).
Compression ratio:	TR20 - 7:1. TS20 - 9:1.
Camshaft:	Sports - 'R' type. Valve Clearances: Inlet 0.002in. Exhaust 0.004in, (cold).
Ignition type:	Lucas Energy Transfer. RM 19 Alternator.
Ignition timing:	TR20 - 12° BTDC, (= 32° fully advanced). Rotor on keyway no. 2 (7 o'clock).
	TS20 - 20° BTDC, (= 40° fully advanced). Rotor on keyway no. 1 (3 o'clock).
Points gap:	0.014in - 0.016in.
Spark plug:	TR20 - Champion L7. TS20 - KLG F80, Lodge 2HN. Gap 0.020in.
Carburettor:	TR20 - Monobloc 375/44, $^{25}/_{32}$in choke size, 'D' shaped air filter.
	TS20 - Monobloc 376/272, $^{15}/_{16}$in choke size, 'D' shaped air filter.
Gearbox ratios:	TR20 - Wide (WR), ratios. TS20 - Extra Close ratios, (ECR).

		1st - All	3rd - WR ECR	2nd - All	4th - WR ECR
	Mainshaft	16T	25T 23T	20T	29T 25T
	Layshaft -	29T	20T 22T	25T	16T 20T

Wheel & Tyre sizes:	Front.	Both models - 2.75 x 21 on WM1 rim.
	Rear.	TR20 - 4.00 x 18 on WM3 rim.
		TS20 - 3.50 x 19 on WM2 rim.
	Tyres.	Dunlop Trials on TR20. Dunlop 'Knobblies' on TS20.
Sprockets & Chains:	Engine 19T.	Clutch 48T. Primary chain $^3/_8$in x $^7/_{32}$in x 62 links, duplex.
	Gearbox 16T.	Rear wheel 58T. Rear chain $^1/_2$in x $^3/_{16}$in x 119 links.
Capacities:	Petrol.	$2^5/_8$ gallons (= 11.92 litres). T15 'Teardrop' shape tank.
	Oil.	$2^3/_4$ pints (= 1.56 litres). Froth tower on tank.
	Gearbox.	$^1/_3$ pint (= 200cc).
	Primary Chain.	$^1/_3$ pint (= 200cc).
	Front fork.	$^1/_4$ pint (= 150cc) 1962/4.
		$^1/_3$ pint (= 200cc) 1965.
Exhaust system:		High level pipe running inside the rear sub-frame.
		The TR20 had a silencer, (angle cut at the end).
		The TS20 used a plain pipe with or without extension.
Colours:	1962/3 -	Burgundy & Silver Sheen.
	1964/5 -	Hi Fi Scarlet and Silver Sheen.
	Petrol Tank	top half - Burgundy or Hi Fi Scarlet.
		lower half - Silver Sheen.
	Mudguards -	Polished alloy.
	Oil Tank -	Black.
	Front Brakeplate -	Silver Sheen (alternatively black for 1962/3).
	Tank Badges -	Black lettering on gold and chrome backing.
	Other painted parts -	Black.
Seat:		Short black single saddle with parallel sided springs.

Notes:

The TR20 and TS20 were specialist trials and scrambles developments of the T20SS model. Introduced in February 1962, they were described as 'Works Replica' models, and were as near as could be bought to the machines used by factory riders.

Both models had upswept exhaust sytems, crankcase undershields, short competition seats, polished alloy mudguards and no lights, toolbox, prop or centre stands. They had heavyweight front forks without top covers, wide handlebars (braced in the case of the TS20), with ball-ended levers, a shortened left footrest bracket carrying the rear brake pedal and sealed wheel bearings. Other equipment included a bulb horn on the TR20 (none on the TS20), a 'D' shaped speedometer for the TR20 only, rearward facing footrests on the TR20 and normal footrests on the TS20. A rear number plate and competition front number plate were available. Also available were a large range of gearbox and rear wheel sprockets. The models remained largely unchanged during their existence except for improvements made to all Cub models over the years. The rear sub-frame had, on the right hand side, a tube cranked outwards at the top and bottom to allow the exhaust pipe to fit inside the sub-frame. Adjustable rear suspension units were available for 1963/4. From the start the front frame loop was double gusseted under the headstock. In 1965 the front forks were changed to a type with the springs outside the stanchions. The small Terrier type petrol tank was used and there was a large air cleaner. Ignition was by Energy Transfer. The 1962 model had a distributor engine but later machines were of the side points type. The first machines were despatched from Meriden in April 1962 and the last in March 1965. Some early engines are numbered T20SST no. rather than TR20 no.

The only reference to these models in any parts list is as a supplement to Parts List No. 8.

			TR20	TS20
Machines built -	1962 - no. 81890 to no. 88346 =		270	24
	1963 - no. 88347 to no. 94599 =		68	340
	1964 - no. 94600 to no. 99732 =		100	1
	1965 - no. 99733 to no. 100013 and no. 101 on =		100	2
	Total number built =		538	367

Analysis of sales:	TR20	1962 -	UK = 74.1%	USA = 21.9%	Others = 4.0%
		1963 -	UK = 82.4%	USA = None	Others = 17.6%
		1964 -	UK = 85.0%	USA = None	Others = 15.0%
		1965 -	UK = 98.0%	USA = None	Others = 2.0%.
	TS20	1962 -	UK = 66.7%	USA = None	Others = 33.3%.
		1963 -	UK = 0.6%	USA = 96.8%	Others = 2.6%.
		1964 -	UK = None	USA = None	Others = 100%.
		1965 -	UK = None	USA = None	Others = 100%.

A 1965 T20SM Mountain Cub in Hunting Yellow. Note the splined type kick-start lever and the first use of externally sprung heavy-weight front forks. (The Mike Estall Collection)

A 1967 T20M Mountain Cub from the sales brochure. (The Mike Estall Collection)

Appendix 1.19 Model profile

T20SM & T20M Mountain Cubs - 1964 to 1967

Machines T20SM 95152 (October 1963) - T20M 9167 (April 1967).

Bore & Stroke:	63mm (2.480in) x 64mm (2.520in).				
Compression ratio:	7:1.				
Camshaft:	Sports - 'R' type. Valve clearances - Inlet 0.002in. Exhaust 0.004in. (cold).				
Ignition type:	Lucas Energy Transfer. RM 19 alternator.				
Ignition timing:	16° BTDC, (= 36° fully advanced). Rotor on keyway no. 1 (3 o'clock).				
Points gap:	0.014in - 0.016in.				
Spark plug:	Champion L7. Gap = 0.020in.				
Carburettor:	Monobloc 376/314, $^{15}/_{16}$in choke size, with pancake air filter.				

Gearbox ratios:	Wide ratios.	1st	3rd	2nd	4th
	Mainshaft	16T	25T	20T	29T
	Layshaft	29T	20T	25T	16T

Wheel & Tyre sizes:	Front.	3.00 x 19 on WM1 rim.	
	Rear.	4.00 x 18 on WM3 rim.	
	Tyres.	Dunlop Trials Universal.	
Sprockets & Chains:	Engine 19T.	Clutch 48T.	Primary chain.$^{3}/_{8}$in x $^{7}/_{32}$in x 62 links, duplex.
	Gearbox 17T.	Rear wheel 54T.	Rear chain $^{1}/_{2}$in x $^{3}/_{16}$in x 117 links.
Capacities:	Petrol.	2$^{5}/_{8}$ gallons, (= 11.92 litres). T15 'Teardrop' shape tank.	
	Oil.	2$^{3}/_{4}$ pints (= 1.56 litres). Froth tower on tank.	
	Gearbox.	$^{1}/_{3}$ pint (= 200cc).	
	Primary Chain.	$^{1}/_{3}$ pint (= 200cc).	
	Front fork.	$^{1}/_{4}$ pint (= 150cc) 1964.	
		$^{1}/_{3}$ pint (= 200cc) 1965/7.	

Exhaust system:	High level pipe running inside the rear sub-frame, with silencer.

Colours:	1964 -	First colour - Crystal Blue. Second colour - Silver Sheen.
	1965 -	Hunting Yellow.
	1966/7 -	First colour - Grenadier Red. Second colour - Alaskan White.
	Petrol Tank:	
	(1964, 1966/7) -	top half - First colour.
		lower half - Second colour.
	(1965) -	whole tank - Hunting Yellow.
	Mudguards:	
	(1964/5) -	Polished alloy.
	(1966/7) -	Alaskan White, sometimes with Grenadier Red centre stripe.
		Painted alloy at the front. Steel at the rear.
		Polished alloy option also available for 1966/7.
	Oil Tank & Toolbox -	Black.
	Front Brakeplate:	
	(1964/5) -	Black.
	(1966/7) -	Black or Silver Sheen.
	Other painted parts -	Black.
	Tank Badges:	
	(1964/5) -	Black lettering on gold and chrome.
	(1966/7) -	White lettering on Grenadier Red and chrome.

Twinseat:	Two level. Grey top with black sides and grey lower styling band. Black version also seen. Safety strap option. 1967 had the 'Triumph' logo in gold at the rear of the seat.

Notes:
The Mountain Cubs were offered as 'trail' bikes, mostly to the American market. It is believed that the bright colours were intended to make machine and rider stand out from the surroundings so that they were less likely to become the targets of hunting parties out shooting game!
The model was developed from the T20SS and T20SC and the specification made the machine suitable for mixed on and off-road use. They were fitted with the Terrier type tank, Energy Transfer ignition and lights, a speedometer (with tachometer option), wide ratio gears, trials tyres, upswept handlebars with ball-ended levers, a crankcase undershield, a small triangular shaped toolbox and folding footrests. The T20SM had a centre stand and pillion footrests but the T20M had neither. The rear light unit was a Lucas L564 for 1964 and L679 for 1965 onwards. The machines had heavyweight front forks, the 1964 T20SM with internal springs and all later machines having springs outside the stanchions. Adjustable rear suspension units were optional. 1965 models had the cranked and pivoted type of kickstart lever. Head and barrel type were 'Oval' in 1964, then 'Square' from February 1965. These machines were not separate models but merely different model descriptions, with only minor alterations to the machine specification. The Mountain Cub started as the model T20SM in 1964, but by September 1965 the model type T20M was being used in sales literature; however, the first T20M numbers were not stamped onto frames and engines until late 1966. The T20SM description also continued to be used long after that date. The figures below include the French Army Cub machines. Due to the inadequacy of the extant records, it has not been possible to give more detail than is shown below.

Parts Lists - The T20SM is covered in Parts List No.1 (no. 101 onwards).

Mountain Cubs built:	T20SM 1964 - no. 94600 to no. 99732 =	801
	T20SM 1965/6 - no. 99733 to no. 100013 and no. 101 & no. 2001 on =	3220
	T20SM & T20M 1967 - no. 2001 & no. 3001 on =	2106
	Total number built =	6127

Analysis of sales:	T20SM	1964	UK = None	USA = 99.4%	Others = 0.6%
	T20SM	1965/6	UK = 1.4%	USA = 80.0%	Others = 18.6%
	T20SM & T20M	1967	UK = 5.6%	USA = 58.4%	Others = 36.0%

A T20M.WD French Army Cub. The machines were fitted with panniers, crashbar, and a handlebar mirror in France. (The Mike Estall Collection)

A T20M.WD French Army Cub from the left side. (The Mike Estall Collection)

Appendix 1.20 Model profile

T20M.WD French Army - 1964 to 1967

Machines T20SM 99639 (July 1964) - T20M 10050 (July 1967).

Bore & Stroke:	63mm (2.480in) x 64mm (2.520in).				
Compression ratio:	7:1				
Camshaft:	Standard type. Valve clearances - Both 0.010in (cold).				
Ignition type:	Lucas Energy Transfer. RM 19 alternator.				
Ignition timing:	22º BTDC, = 32º fully advanced. Rotor on keyway no. 1 (3 o'clock).				
Points gap:	0.014in - 0.016in.				
Spark plug:	Champion L7. Gap 0.020in.				
Carburettor:	Monobloc 375/44, $^{25}/_{32}$in, with 'D' shaped air filter.				
Gearbox ratios:	Wide ratios.	1st	3rd	2nd	4th
	Mainshaft	16T	25T	20T	29T
	Layshaft -	29T	20T	25T	16T
Wheel & Tyre sizes:	Front.	3.00 x 19 on WM1 rim.			
	Rear.	3.50 x 18 on WM2 rim.			
	Tyres.	Dunlop Trials.			
Sprockets & Chains:	Engine 19T.	Clutch 48T.	Primary chain $^3/_8$in x $^7/_{32}$in x 62 links, duplex.		
	Gearbox 17T.	Rear wheel 58T.	Rear chain $^1/_2$in x $^5/_{16}$in x 119 links.		
Capacities:	Petrol.	$2^5/_8$ gallons (= 11.92 litres). T15 'Teardrop' shape tank.			
	Oil.	$2^3/_4$ pints (= 1.56 litres). Froth tower on tank.			
	Gearbox.	$^1/_3$ pint (= 200cc).			
	Primary Chain.	$^1/_3$ pint (= 200cc).			
	Front fork.	$^1/_4$ pint (= 150cc), 1964.			
		$^1/_3$ pint (= 200cc) 1965/7.			
Exhaust system:	Low level.				
Colours:	All painted parts of the machine, (except rear number plate), were NATO anti infra-red green, BSC 298.				
	The tank badges had white lettering on a NATO green and chrome backing strip.				
Twinseat:	All black.				

Notes:
The French Army Cub was a side points engined T20T with Mountain Cub cycle parts, strengthened and modified to the requirements of the French Army. It had the small Terrier type tank, heavyweight front forks and non-adjustable rear suspension. The front frame was equipped with double gusseting under the headstock and the rear sub-frame had a high level exhaust mounting bracket on the right hand side. This tube was straight, i.e. was not cranked at the top or bottom. However, a low level exhaust system was used. The swinging arm had a bracket on the left side to provide an additional mounting point for the large chainguard. There was a crankcase undershield, reinforced, non-folding footrests, a small triangular toolbox and sealed wheel bearings. The machine had upswept handlebars with ball-ended levers. There was a new oil tank with the feed and return pipes sited further to the rear to prevent them rubbing on the crankcase. The machine was usually equipped with panniers, a handlebar mirror and crashbars once it had arrived in France.
Mechanically the engine unit had Energy Transfer ignition, a 7:1 piston modified to take the 9:1 piston rings which had a greater radial depth, the standard cam, sports valve springs, a large diameter needle roller big end, the square cylinder head and barrel, a slider block driven oil pump and wide ratio gears. The rocker feed pipe was heavier than standard, as was the rear chain which was $^1/_2$in x $^5/_{16}$in instead of the usual $^1/_2$in x $^3/_{16}$in. There may have been both 'road' and 'off-road' versions. The model appears to have suffered frequent problems with oil leaks behind the clutch, frame breakages under the headstock and near the front engine mount, ignition system failures and oil leaks.
The numbers built will probably never be known with certainty due to the poor records now remaining in existence. It is known that the number originally ordered was 7,000 machines but it is also known that this order was never fulfilled. Detailed research has accounted for 1,480 machine serial numbers. Apart from the first 82 machines in 1964 all the French Army Cubs were built at the BSA Small Heath factory up to July 1967. However, many went into storage and deliveries seem to have continued right into mid-1970.

Parts Lists - There is a Parts List for the French Army Cub, a specification leaflet and a Riders Manual, all in French.

Analysis of sales (all to CGCIM, France):

1964 =	82	
1965 =	Nil	
1966 =	722	
1967 =	303	
1968 =	100	
1969 =	144	(Plus 100 T20B Super Cubs).
1970 =	51	
Total =	1402	(Plus an unknown number, currently 78, not shown in the extant BSA records).

An artist's impression of the T20B Bantam Cub. The exhaust entry is now outboard of the silencer centreline. (The Mike Estall Collection)

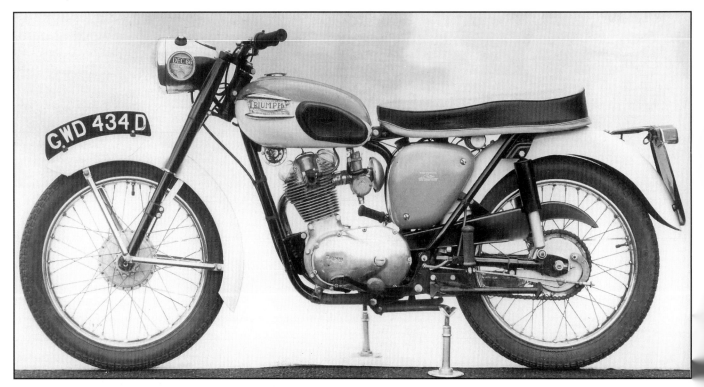

A T20B Bantam Cub from the drive side. A mixture of Bantam D7 cycle parts and Cub engine. (Courtesy EMAP)

Appendix 1.21 Model profile

T20B Bantam Cub - 1966 to 1968

Machines T20B 101 (December 1965) - T20B 7849? (March 1967).

Bore & Stroke:	63mm (2.480in) x 64mm (2.520in).
Compression ratio:	7:1.
Camshaft:	Standard. Valve clearances - both 0.010in, (cold).
Ignition type:	Lucas* AC/DC with battery. RM 18 alternator. * Wipac on some machines.
Ignition timing:	8° BTDC, = 32° fully advanced. Rotor on keyway no. 1, (3 o'clock).
Points gap:	0.014in - 0.016in.
Spark Plug:	Champion L7, Lodge HN, KLG F80. Gap 0.025in.
Carburettor:	Monobloc 375/61, $^{25}/_{32}$in, with screw-on pancake air filter.

Gearbox ratios:

	Standard ratios.	1st	3rd	2nd	4th
	Mainshaft -	16T	25T	20T	28T
	Layshaft -	29T	20T	25T	17T

Wheel & Tyre sizes:	Both.	3.00 x 18 on WM1 rim.
	Tyres.	Dunlop. Block pattern front and rear.

Sprockets & Chains:	Engine 19T.	Clutch 48T.	Primary chain $^3/_8$in x $^7/_{32}$in x 62 links, duplex.
	Gearbox 17T.	Rear wheel 47T.	Rear chain $^1/_2$in x $^3/_{16}$in x 115 links.

Capacities:	Petrol.	$2^1/_4$ galls (= 10.25 litres). One-piece tank badges.
	Oil.	4 pints (= 2.27 litres).
	Gearbox.	$^1/_3$ pint (= 200cc).
	Primary Chain.	$^1/_3$ pint (= 200cc).
	Front fork.	$^1/_8$ pint (= 75cc).

Exhaust system:	Low level.

Colours: Nutley Blue and Alaskan White. n.b. - Nutley Blue was also shown in some literature as Perrivale Blue. (Some sales literature also shows Pacific Blue - but the colour was changed after the brochure was printed).

Petrol Tank -	top half - Nutley Blue.
	lower half - Alaskan White.
Mudguards -	Alaskan White, with gold pin-stripes around the central rib on the front mudguard.
	Plain section rear mudguard.
Oil Tank & Toolbox -	Nutley Blue.
Brakeplates -	Silver, front and rear.
Hubs -	Silver, front and rear. (Single sided hubs).
Tank Badges -	Blue lettering on gold and chrome backing.
Other painted parts -	Black.
Headlamp Shell -	Chrome plating.

Twinseat:	D7 Bantam type. Grey or black top, with or without safety strap.

Notes:

The Bantam Cub was announced in December 1965. It used D7 Bantam cycle parts with an oil tank and the frame modified by the addition of a third engine mounting point for the Cub engine lug underneath the crankcase. The first machine was numbered T20B 101.

The engine was the latest standard road Cub type - side points, initially a plain big end but from August 1966 a single row caged needle roller was used, and an enlarged sludge trap. It had two ball races for the main bearings, and the 'Square' head and barrel. It was for the most part fitted with the latest oil pump drive - a slider block driven type, similar to that used on the twin cylinder machines, giving a longer stroke to the oil pump and a 50% increase in the delivery of oil, but early machines had the older eccentric pin driven pump. The engine was fitted with a single high gear oil seal.

Electrics were by Lucas except for a Wipac dip/horn button and stoplight switch, but there were some machines built with a Wipac wiring loom, (harness number 54138). Ball-ended handlebar levers were used. The brake drums were 5$^1/_2$in single sided on 18in rims (not to be confused with the full width hubs used on the later Super Cub). The front forks were the same internally as the lightweight Cub forks but the stanchions were 1$^5/_8$in longer and fitted with conventional top covers and a separate headlamp shell, rather than the Tiger Cub nacelle. The exhaust pipe and silencer were new items and different to the T20. (The pipe had an outward kink in the lower run and the silencer entry pipe was outboard of the silencer body centre line, whereas the T20 pipe was straight and the silencer entry pipe was inboard of the centre line.) There was no steering lock.

For road test figures see Appendix 8.

The BSA numbering method was such that the same number was used for different models several times over. The number was always preceded by a prefix denoting the model type. However, this method was not infallible as there were repetitions. The T20B series was used for both the Bantam Cub and the Super Cub. The numbers themselves were not used in sequence so it cannot be said that, because one machine has a higher number than another, it is a later machine!

There is a Parts List for the Bantam and Super Cubs.

The majority of Bantam Cubs were built in 1966 but small batches continued to be built up to June 1967. Because the records are inadequate it is often not possible to say with certainty which machines were Bantam Cubs and which were Super Cubs, or in which season they were built. Therefore some 'intelligent' guesses have had to be made in the analysis below. The serial number of the last Bantam Cub, shown in the header, is also uncertain.

Best estimate of the number of T20B Bantam Cub machines built = 1,719

Analysis of sales: UK = 54.6%   USA = Nil   Others = 45.4%

A T20B Super Cub. Like the Bantam Cub it was a hybrid model: this time a D10 Bantam with a Cub engine. (Courtesy EMAP)

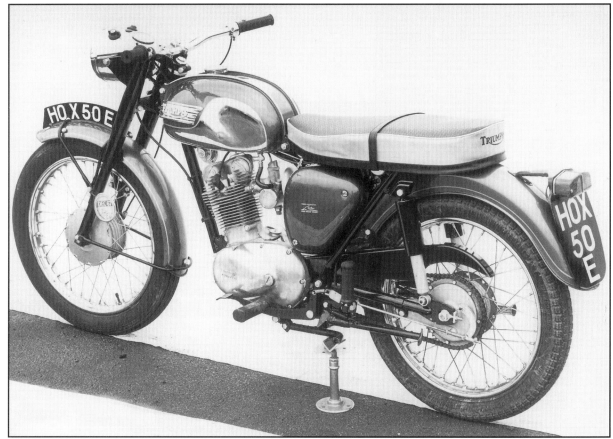

An artist's impression of the T20B Super Cub from the drive side. (Courtesy EMAP)

Appendix 1.22 Model profile

T20B Super Cub - 1967 to 1969

Machines T20B 3333? (October 1966) - T20B 9710? (June 1969).

Bore & Stroke:	63mm (2.480in) x 64mm (2.520in).				
Compression ratio:	7:1.				
Camshaft:	Standard. Valve clearances - both 0.010in, (cold).				
Ignition type:	Lucas* AC/DC with battery. RM 18 alternator. * Wipac on some machines.				
Ignition timing:	8° BTDC, = 32° fully advanced. Rotor on keyway no. 1 (3 o'clock).				
Points gap:	0.014in - 0.016in.				
Spark plug:	Champion L7, Lodge HN, KLG F80. Gap 0.025in.				
Carburettor:	1967 - Monobloc 375/61, $^{25}/_{32}$in choke size.				
	1968/9 - 22mm Concentric R622/1. Pancake air filter.				
Gearbox ratios:	Standard ratios.	1st	3rd	2nd	4th
	Mainshaft -	16T	25T	20T	28T
	Layshaft -	29T	20T	25T	17T
Wheel & Tyre sizes:	Both.	3.00 x 18 on WM1 rim.			
	Tyres.	Dunlop. Block pattern front and rear.			
Sprockets & Chains:	Engine 19T.	Clutch 48T.	Primary chain $^3/_8$in x $^7/_{32}$in x 62 links, duplex.		
	Gearbox 17T.	Rear wheel 47T.	Rear chain $^1/_2$in x $^3/_{16}$in x 115 links.		
Capacities:	Petrol.	2 galls (= 9.13 litres). Two-piece tank badges.			
	Oil.	4 pints (= 2.27 litres).			
	Gearbox.	$^1/_3$ pint (= 200cc).			
	Primary Chain.	$^1/_3$ pint (= 200cc).			
	Front fork.	$^1/_8$ pint (= 75cc).			
Exhaust system:	Low level.				
Colours:	1967 -	Bushfire Red*, black & chrome. White lining.			
	1968/9 -	Firecracker Red*, black & chrome. Gold lining.			
		*These are thought to be different names for the same colour.			
	Petrol Tank -	Bushfire/Firecracker Red on chrome with white or gold lining. Chrome styling strips over the seams.			
	Mudguards -	Bushfire/Firecracker Red, black centre stripe lined in white or gold.			
	Oil Tank & Toolbox -	Bushfire or Firecracker Red.			
	Hubs & Brakeplates -	Silver, front and rear. (Full-width hubs).			
	Tank Badges -	Black lettering on gold and chrome backing.			
	Other painted parts -	Black.			
	Headlamp Shell -	Chrome plating.			
	Cylinder -	The last Super Cubs may have had black painted barrels.			
Twinseat colour:	All black or grey sides with black top or two shades of grey, usually with safety strap. The top was sometimes diamond quilted.				

Notes:
The Super Cub was announced for the 1967 season. It was essentially D10 (which later became the D14) Bantam cycle parts with the frame modified by the addition of a third engine mounting point to accommodate the Cub engine. It was sometimes referred to as the 'T20 De Luxe'. The first machines were despatched in December 1966.
The engine was the same as in the Bantam Cub introduced a year earlier. It was the latest standard road Cub type - side points, a flywheel assembly with a needle-roller big end and the enlarged sludge trap. It had two ball races for the main bearings, and the square head and barrel. It was also fitted with the slider block drive for the oil pump, giving a longer stroke to the oil pump and thereby increased pumping capacity.
Electrics were initially by Lucas but from January 1967 most machines used a Wipac wiring harness, a Wipac S.3611 rear light unit and a Wipac Ducon horn/dipper switch. Ball-ended bolt-on handlebar levers were used. The brake drums were $5^1/_2$in full width on 18in rims (not to be confused with the single sided hubs used on the earlier Bantam Cub). The front forks were the same internally as the lightweight Cub forks but the stanchions were $1^5/_8$in longer and were fitted with conventional top covers and a separate headlamp shell, rather than the Tiger Cub nacelle. The exhaust pipe and silencer were new items and different to the T20. (The pipe had an outward kink in the lower run and the silencer entry pipe was outboard of the silencer body centre line, whereas the T20 pipe was straight and the silencer entry pipe was inboard of the centre line.) There was no steering lock.

For road test figures see Appendix 8.

The BSA numbering method was such that the same number was used for different models, several times over. The number was always preceded by a prefix denoting the model type which in this case was T20B (the same as used earlier for the Bantam Cub). However, this method was not infallible as there were repetitions. The T20B series was used for both the Bantam Cub and the Super Cub. The numbers themselves were not used in sequence so it cannot be said that, because one machine has a higher number than another, it is a later machine!

There is a Parts List for the Bantam and Super Cubs.

The majority of Super Cubs were built in the 1967 season (*i.e.*, from introduction at the November 1966 Earls Court Show until the end of July 1967), but limited production continued through to June 1969. Because the records are inadequate it is often not possible to say with certainty which machines were Bantam Cubs and which were Super Cubs, or in which season they were built. Therefore, some 'intelligent' guesses have had to be made in the analysis below. The serial number of both the first and last Super Cub is uncertain.

Best estimate of the number of T20B Super Cub machines built = 2,450

Analysis of sales: UK = 58.4%   USA = 1 machine   Others = 41.6%

Appendix 2

Production year[1] - first and last serial numbers

(From the Meriden and Small Heath despatch and build books - but see Notes 8 & 9)

Season	1954	1955	1956	1957	1958	1959	1960	1961	1962	1963	1964	1965	1966 - onwards
Start nos.	101	8518	17389	26276	35847	45312	56360	69517	81890	88347	94600	99720 101 2001	101 2001 3001
Finish nos.	8517	17388	26275	35846	45311	56359	69516	81889	88346	94599	99719	100013 Various Various	Various Various Various

Model types

(First and last serial numbers and approximate 'Build' dates of each model type)

Season[1]	1954	1955	1956	1957	1958	1959	1960	1961	1962	1963	1964	1965	1966	1967/69
T15	101 7/53	+	+	31463 4/57										
T20	3000 3/54	+	+	+	+	+	+	+	+	+	+	+	4011 11/65	
T20 150[2]			24806 5/56	+	+	+	+	+	+	+	98690 5/64			
T20C				27660 1/57	+	52974 5/59								
T20C 150[2]					32317 5/57	50394 2/59								
T20J					36365 10/57	+	+	70805 10/60						
T20CA[3]					39773 2/58	53780 6/59								
T20S						48139 12/58	+	75973 3/61						
T20T						56145 9/59	+	+	85698 5/62			833 11/64		
T20W							60265 60290 1/60							
T20SL[4]							64592 4/60	81889 7/61						
T20J 150[2]								70804 70805 10/60						
T20SS									82276 9/61	+	+	1715 1/65		
T20SH									82756 10/61	+	+	+	4373 1/66	
T20SC									84912 3/62	+	+	2034 7/65		

Season[1]	1954	1955	1956	1957	1958	1959	1960	1961	1962	1963	1964	1965	1966	1967/69
T20SR									84967 3/62	+	+	2029 7/65		
TR20									85108 3/62	+	+	1867 2/65		
TS20									85323 3/62	+	+	1870 2/65		
T20SS 150[2]									85445 4/62	+	95368 10/63			
T20WD[5]										88658 9/62	99719 7/64			
T20SM[7]											95152 10/63	+	+	9167 4/67
French Army[6]											99639 7/64	+	+	10050 7/67
T20P												775 11/64	2619 9/65	
T20B (BantamCub)[7]													101 12/65	to 7849 3/67
T20M[7]													3001 10/66	to 8179 4/67
T20B (SuperCub)[7]													3333 10/66	to 9710 6/69
T20B (Tarbuk)[7]													2422 1/68	to 5668 3/68
T20 Bantam175														Numbers? 1/69 to 5/69

Notes:

1. The 'Season' or 'Production Year' shown was the same as the factory 'Financial Year', which started on 1st August and finished on 31st July.

2. 150[2] - These were T20, T20C, T20J and T20SS models fitted with 150cc engines for the island of Bermuda.

3. There was a proposed developement of the T20CA model in February 1959, the T20CB, but it was overtaken by the trials version of the T20S and it is believed that none were built.

4. T20SL 64592 was delivered in November 1960 (1961 year), but it had a serial number dating from April 1960.

5. T20WD 88658, which had a September 1962 number, was not actually delivered until December 1964.

6. The French Army Cub was shown in the despatch records under various model types or sometimes no model type at all. It appeared as T20WD, T20SM or T20M as well as T20M.WD, or as an unknown model type delivered to CGCIM, Paris. The first machine shown was a T20WD in August 1964 and the last shown was a T20M in May 1970.

7. The numbering of the T20B Bantam Cubs, Super Cubs and 'Tarbuk' machines and the T20SM and T20M Mountain Cubs was ambiguous. The records frequently did not show the model type or sometimes only 'Export' or 'T20B'.

   The earliest delivery date shown for a Super Cub was T20B 4016 in November 1966. However, there were several machines delivered after that date but having lower serial numbers - the lowest being no. 3043 delivered in October 1967. But there also exists a factory memorandum showing no. 3333 as the first Super Cub.

   The highest numbered Super Cub shown in the despatch records was no. 9710, delivered in May 1968, but the last machines were built in June 1969. Their serial numbers are unknown. Apart from one or two 'strays', the last Super Cubs went out from Small Heath in July 1969.

8. There is a great deal of uncertainty about the serial numbers and dates for many of the models built at and delivered from Small Heath due to the inadequacy of the records remaining in existence.

9. The Tiger Cub and Terrier Register now shows the current existence of several machines whose serial numbers lay outside the limits shown above. The dates and serial numbers above show what is to be found in the factory records existing today and should only be treated as best estimates.

Appendix 3

Component dimensions

(All measurements are in inches - unless otherwise shown)

Piston and cylinder:

Bore and Stroke: T15   57.0mm x 58.5mm (2.244 x 2.303).
                T20   63.0mm x 64.0mm (2.480 x 2.520).

Cylinder: Inclined forward 25° from the vertical.

Piston clearance limits:

	T15 (7:1)	T20 (7:1)	T20 (9:1)
Top land -	-0.012 to -0.015	-0.012 to -0.015	-0.012 to -0.018
Centre land -	-0.010 to -0.012	-0.010 to -0.012	-0.011 to -0.013
Bottom land -	-0.010 to -0.012	-0.010 to -0.012	-0.011 to -0.013
Top skirt* -	-0.0050 to -0.0055	-0.0055 to -0.0060	-0.0070 to -0.0075
Bottom skirt* -	-0.0035 to -0.0040	-0.0015 to -0.0020	-0.0030 to -0.0035
Ovality of skirt -	No information	0.006	0.010
*Fore & aft dimension.			

Piston diameter at top of skirt: No information   2.4745 to 2.4740   No information

Compression height:

To gudgeon pin centre -	No information	1.120	1.430
Piston length -	No information	2.0555	2.3650

Piston oversizes: T15   Std. =   2.244 = 57.00mm. Gives a capacity of 149.27cc.
                       +20 =   2.264 = 57.50mm. Gives a capacity of 151.91cc.
                       +40 =   2.284 = 58.01mm. Gives a capacity of 154.61cc.

              T20   Std. =   2.480 = 63.00mm. Gives a capacity of 199.50cc.
                       +10 =   2.490 = 63.25mm. Gives a capacity of 201.09cc.
                       +20 =   2.500 = 63.50mm. Gives a capacity of 202.68cc.
                       +30 =   2.510 = 63.76mm. Gives a capacity of 204.35cc.
                       +40 =   2.520 = 64.01mm. Gives a capacity of 205.95cc.
                       +60 =   2.540 = 64.52mm. Gives a capacity of 209.25cc.
                       +80 =   2.560 = 65.03mm. Gives a capacity of 212.57cc. (USA 10:1 Robbins 'slipper' piston).

Piston ring gap: T15 = 0.006 to 0.008.
                     T20 = 0.008 to 0.010.

Piston ring depth: $^1/_{16}$ compression rings.
                     $^3/_{32}$ scraper ring.

Piston ring radial thickness: 7:1 piston - 0.091 maximum.
                               9:1 piston - 0.105 maximum.

Small end (nominal): $^9/_{16}$ bore x $^3/_4$ long.
Small end bush dimensions -   O.D. - 0.6865 to 0.6875
                                I.D. - 0.5625 to 0.5630

Gudgeon pin diameter: 0.5617 to 0.5625 (= $^9/_{16}$ nominal).
Clearance in small end bush: 0.001

Cylinder barrel, studs & nuts:

	Barrel type	Stud length	Nut height
Terrier	'Round' E3142	'Short' 4$^{19}/_{32}$ E3170	$^9/_{16}$ E3230
Cubs from no. 3000	'Round' *E3385	'Long' 4$^7/_8$ E3365	$^9/_{16}$ E3230
Cubs from no. 45312	'Oval' *E3385	'Long' 4$^7/_8$ E3365	$^9/_{16}$ E3230
Cubs from no. 56360	'Oval' E3385	'Short' 4$^{19}/_{32}$ E3170	$^{11}/_{16}$ E4103
Cubs from no. 101	'Square' E6094	'Short' 4$^{19}/_{32}$ E3170	1$^1/_{16}$ E5897

*Cub 'Round' and 'Oval' barrels all have the same part number.

Crankshaft:

Big end types-

	Big end - type	Big end - size	Crankpin spigot size
T15 from no. 101-	15 rollers single row Crankpin diameter = 0.962	$1^1/_2$ (OD)	$^3/_4$ diameter = 0.754.
T15 from no. 3905 -	Plain bush	$1^1/_8$ bore x $^5/_8$ long	$^3/_4$ diameter = 0.754.
T20 models from no. 3000 -	Plain bush	$1^1/_8$ bore x $^5/_8$ long	$^3/_4$ diameter = 0.754.
Also -	14 rollers, $^1/_2$ x 0.136, steel cage.	1.141 conrod eye diameter. Crankpin diameter = 0.864.	$^3/_4$ diameter = 0.754.
Also -	10 rollers, $^5/_{32}$ diameter.	1.129 conrod eye diameter. Crankpin diameter = 0.754.	$^3/_4$ diameter = 0.754.
All models from no. 17389* -	Plain bush	$1^1/_8$ bore x $^5/_8$ long	$^{13}/_{16}$ diameter = 0.8153.
All models from no. 84269 -	Plain bush	$1^5/_{16}$ bore x $^5/_8$ long	$^{13}/_{16}$ diameter = 0.8153.
All models from no. various - Aug.1966	Caged needle roller, double row		

*Flywheel rim width increased from no. 18597

Conrod dimensions

T20 models from no. 3000 -	Crankpin track diameter =	1.1225 to 1.1230
T20 models from no. 3000 -	Conrod big end eye diameter =	1.245 to 1.250
T20 models from no. 84269 -	Crankpin track diameter =	1.3102 to 1.3105
T20 models from no. 84269 -	Conrod big end eye diameter =	1.3125 to 1.3135

The crankpin is a 0.004 interference fit into the flywheels on all machines.
Big end (plain bush type) maximum permissible vertical clearance = 0.0015 to 0.0020.
Conrod centres (big end to small end) - all models = $4^7/_8$.

Mainshafts and keyways:

Permissible 'run-out' on drive side mainshaft = 0.001.

Drive side mainshaft keyways - Use no. 1 keyway (@ 3 o'clock) for all models with AC/DC ignition.
Use no. 1 keyway (@ 3 o'clock) for Energy Transfer models with high compression pistons.
Use no. 1 keyway (@ 3 o'clock) for T20SM, T20M and Energy Transfer French Army machines.
Use no. 2 keyway (@ 7 o'clock) for other Energy Transfer models with low compression pistons.

Timing pinion bolt (up to engine no. 84268) - maximum permitted length = $1^{29}/_{32}$.

Main bearings:

Timing side:
Up to no. 84268, a plain bush:

Main bearing bush - nominal size =	$1^1/_8$ bore x $^{15}/_{16}$ long.
Main bearing bush - internal diameter =	1.1260 to 1.1265 - (before fitting).
Main bearing bush - internal diameter =	1.1250 to 1.1255 - (after fitting).
Main bearing bush - bush housing in crankcase =	1.4380 to 1.4385 diameter.
Flywheel journal size =	1.1235.
Clearance between bush and journal =	0.0010 - 0.0015.
Diameter of journal when replacement is due =	1.121.
From engine no. 84269:	6304 ballrace = 52mm OD x 20mm ID x 15mm width.

Drive side:
All engines - 6304 ballrace.

Drive side inner cover (the DS main bearing carrier on engines up to no. 57616) was a 0.0027 interference fit in the crankcase.

Ignition timing:

Model	Piston position BTDC	Crankshaft	Auto unit	Full advance
T15, T20 1954-1955	0.016 = $^1/_{64}$= 0.40mm	8°	12° (x2 = 24°)	32°
T20 1956-1962, T20C, T20CA	TDC	4°	12° (x2 = 24°)	28°
T20 1963 on, T20B	0.016 = $^1/_{64}$ = 0.40mm	8°	12° (x2 = 24°)	32°
T20S, low compression, T20T	0.016 = $^1/_{64}$ = 0.40mm	8°	12° (x2 = 24°)	32°
T20S, high compression, T20SL	0.062 = $^1/_{16}$ = 1.58mm	16°	12° (x2 = 24°)	40°
T20SS, T20SC, T20SR	0.094 = $^3/_{32}$ = 2.38mm	20°	10° (x2 = 20°)	40°
T20SH	0.062 = $^1/_{16}$ = 1.58mm	16°	12° (x2 = 24°)	40°
TR20	0.032 = $^1/_{32}$ = 0.81mm	12°	10° (x2 = 20°)	32°
TS20	0.094 = $^3/_{32}$ = 2.38mm	20°	10° (x2 = 20°)	40°
T20SM, T20M	0.062 = $^1/_{16}$ = 1.58mm	16°	10° (x2 = 20°)	36°
French Army	0.109 = $^7/_{64}$ = 2.78mm	22°	5° (x2 = 10°)	32°

Valve Gear:

Valve dimensions:

Valve head diameter -	Up to engine no. 48138 -	$1^1/_8$, inlet and exhaust. Exhaust head size never changed.
	From engine no. 48139 -	$1^5/_{16}$ inlet on some models, e.g. T20S.
	From engine no. 56360 -	$1^5/_{16}$ inlet on all models.

Valve stem diameter -	Inlet -	0.3095 to 0.3100.
	Exhaust -	0.3090 to 0.3095.

Valve seat angle -	45°.
Valve seat width -	$^3/_{64}$
Valve included angle -	75°.

Valve guides:

Valve guide length -	Up to engine no. 45311 -	1.838 to 1.848. There is $^{41}/_{64}$ above the circlip.
Valve guide length -	From engine no. 45312 -	1.760 to 1.770. There is $^9/_{16}$ above the circlip.
Valve guide bore -		0.3120 to 0.3130.

Valve springs:

	Inner	Outer
Free length -	1.5625, ($1^9/_{16}$).	1.625, ($1^5/_8$).
Fitted length, Standard cam -	1.1880.	1.219.
Fitted length, Sports 'R' cam -	1.160.	1.230.
Number of turns -	7.	$6^1/_2$.
Compressibilty, Standard spring -	6.071 Kg/cm.	14.286 Kg/cm.
Compression force, in situ, valve closed -	5.897 Kg.	14.742 Kg.

Pushrods:

	T15	T20
Pushrod length -	$6^3/_{16}$.	$6^7/_{16}$.
Pushrod tunnel length -	$5^1/_2$.	$5^3/_4$.

Tappets:

Tappet diameter -	0.3110 to 0.3115.
Tappets - clearance in housing -	0.0005 to 0.0020.

Tappet clearances (engine cold):

	Inlet	Exhaust
Standard cam -	0.010.	0.010.
Sports 'R' cam -	0.002.	0.004.
Big Bear cam -	0.005 (0.004?).	0.005 (0.006?).

Rockers:

Rocker spindle diameter -	0.4355 to 0.4360.
Rocker bush bore -	0.4375 to 0.4380.

Camshaft:

Camshaft bush bore, distributor type -	0.5610 to 0.5620.
Camshaft bush bore, left side, side points type -	0.5610 to 0.5620.
Camshaft bush bore, right side, side points type -	0.6240 to 0.6250.
Diameter of camshaft, left side, side points type -	0.5595 to 0.5600.
Diameter of camshaft, right side, side points type -	0.6225 to 0.6230.

Valve timing:

Cam type -	Standard Cam	'R' Cam	T15 'Kit' Cam	Big Bear Cam	Harman & Collins 'Race' or 'Power'
Tappets set at -	0.015	0.020		0.020	
A. Inlet opens BTDC -	30°	39°	30°	34°	52°
B. Inlet closes ABDC -	50°	61°	60°	56°	82°
C. Exhaust opens BBDC -	55°	65°	60°	60°	82°
D. Exhaust closes ATDC -	25°	35°	30°	30°	52°
Valve overlap = A+D -	55°	74°	60°	64°	104°
Cam lift -	0.248	0.312	0.260	0.310	0.312/0.317
Tappet lift* = Cam lift plus 0.032 at tappets -	0.280	0.344	0.292	0.342	0.344 Inlet 0.349 Exhaust

Valve lift*:

To find the actual valve lift add 10% to make allowance for assymetric rockers.

Oil pump:

Oil pump sizes -

	Bore of chamber		Stroke	Material
	Feed	Scavenge	Both	
From no. 101 -	$^3/_{16}$	$^5/_{16}$	$^3/_{16}$	Brass.
From no. 3905 -	$^1/_4$	$^5/_{16}$	$^3/_{16}$	Brass.
From no. 69517 -	$^5/_{16}$	$^3/_8$	$^3/_{16}$	Brass
From no. 84269 -	$^5/_{16}$	$^3/_8$	$^3/_{16}$	Cast Iron.
From August 1965 -	$^5/_{16}$	$^3/_8$	$^9/_{32}$	Brass - (see Note).

Note: From August 1965, on some machines, there was a change in the oil pump drive to a slider block. This gave an increase in the effective stroke of the pump and therefore a 50% increase in the feed and scavenge rates.

Distributor oil pump shaft and skew gear - vertical float = 0.002 - 0.004.

Oil pressure:

	Idling speed	Fast idle or above
With SAE 30 oil -		
TS main bearing without bleed slot -	50 - 75psi.	100 - 125psi.
TS main bearing with bleed slot -	30psi.	No information.

Clutch:

Clutch springs:
- Free length - $1^{21}/_{32}$ (42mm). Minimum permitted - $1^{17}/_{32}$ (39mm).
- Number of turns - $10^{1}/_{2}$ (after no. 94600).
- Compression in situ - 25.855Kg (after no. 94600).

Clutch driving plates:
- Thickness - $^{1}/_{8}$ = plate thickness of $^{1}/_{16}$, plus two lining thicknesses of $^{1}/_{32}$ each.
- Diameter - Simplex = 5.05.
  - Duplex = 4.95.

Clutch pushrod, length - 7.00 to 7.01.
Clutch pushrod, diameter - 0.1563.

Chains:

Chainlines:
- Primary - from conrod centre to centre of engine sprocket:
  - Simplex chain engines from no. 101 - $2^{3}/_{32}$.
  - Duplex chain engines from no. 35847 - $2^{21}/_{64}$.

- Secondary - conrod centre to gearbox sprocket centre - $1^{1}/_{8}$.

Chain sizes:
- Primary from no. 101 - $^{3}/_{8}$ x $^{7}/_{32}$ x 62 links, simplex.
- Primary from no. 17389 - $^{1}/_{2}$ x $^{3}/_{16}$ x 48 links, simplex.
- Primary from no. 35847 - $^{3}/_{8}$ x $^{7}/_{32}$ x 62 links, duplex.

- Secondary all models except military - $^{1}/_{2}$ x $^{3}/_{16}$ x 112 to 119 links.
- Secondary military models - $^{1}/_{2}$ x $^{5}/_{16}$ x 119 links.

Gearbox:

Basic gearset pattern:

	(Appearance)	(Cluster)		(No dogs)	(Sleeve gear)
	(Ident. letter)	A	B	C	D
Mainshaft	Gear(s)	Low=======3rd		2nd	Top
Layshaft	Gear(s)	Low	3rd	2nd=======Top	
	(Ident. letter)	E	F	G	H
	(Appearance)	(Kickstart gear)	(With Dogs)	(Cluster)	

Total tooth count for each pair of gears - A + E = 45   B + F = 45   C + G = 45   D + H = 45

To calculate internal ratios, multiply the tooth count on each pair of pinions, thus:

1st gear = $E \div A \times D \div H$ 2nd gear = $G \div C \times D \div H$
3rd gear = $F \div B \times D \div H$ Top gear = **Always 1:1**

Factory gearsets:

		Tooth count				Internal Ratios
Standard ratio -	Mainshaft	A	B	C	D	1st = 2.99:1
		16T======25T		20T	28T	2nd = 2.06:1
	Layshaft	29T	20T	25T====17T		3rd = 1.32:1
		E	F	G	H	4th = 1.00:1
Wide ratio -	Mainshaft	A	B	C	D	1st = 3.29:1
		16T======25T		20T	29T	2nd = 2.27:1
	Layshaft	29T	20T	25T====16T		3rd = 1.45:1
		E	F	G	H	4th = 1.00:1
Close ratio -	Mainshaft	A	B	C	D	1st = 2.72:1
		16T======25T		20T	27T	2nd = 1.88:1
	Layshaft	29T	20T	25T====18T		3rd = 1.20:1
		E	F	G	H	4th = 1.00:1
Extra-close ratio -	Mainshaft	A	B	C	D	1st = 2.27:1
		16T======23T		20T	25T	2nd = 1.56:1
	Layshaft	29T	22T	25T=====20T		3rd = 1.20:1
		E	F	G	H	4th = 1.00:1
Ultra-close ratio -	Mainshaft	A	B	C	D	1st = 2.06:1
		17T======23T		20T	25T	2nd = 1.56:1
	Layshaft	28T	22T	25T=====20T		3rd = 1.20:1
		E	F	G	H	4th = 1.00:1

Internal dimensions:

High gear bore, including bush -	0.6260 to 0.6270.
Mainshaft diameter, left side -	0.6230 to 0.6235.
Mainshaft diameter, right side -	0.6100 to 0.6105.
Layshaft low gear bore, incl. bush -	0.6120 to 0.6130.
Layshaft, both sides -	0.6100 to 0.6105.
Other gearbox bushes, bore -	0.6120 to 0.6130.

Gearbox sprocket -

Dimension through the high gear oil seal -	Up to no. 35846 = 1.248.
	From no. 35847 = 1.315.

Sprocket sizes:

Gearbox - up to engine no. 35846 -	13 to 18 teeth.
from engine no. 35847 -	13 to 19 teeth.
Rear wheel -	46T, 47T, 48T, 50T, 52T, 54T, 56T, 58T.

High gear bearing:

Type 6205 - 52mm OD x 25mm ID x 15mm width.

Front forks:

Lightweight type:

Stanchion length - T15/T20 =	$19^1/_4$.
T20C =	$20^1/_4$.
T20B Bantam/Super Cub =	$20^7/_8$.

Stanchion diameter -

The lightweight hydraulically damped forks used from 1957 onwards had stanchions that were 0.010 smaller in diameter than the earlier grease filled type. The actual dimensions are not known.

Slider bush dimensions -

Top bush, bore =	1.068 to 1.069.
Top bush, OD =	1.121 to 1.122.
Bottom bush, OD =	1.118 to 1.119.

Oil capacity (SAE 30) (after no. 26276) = 75cc or $^1/_8$ pint.

Heavyweight type:

Two varieties -	springs INSIDE the stanchions.	Springs OUTSIDE the stanchions.
Oil capacity -	150cc or $^1/_4$ pint.	200cc or $^1/_3$ pint.

Stanchion diameter =	33mm.
Stanchion length =	21in.

Fork Springs:

	Lightweight	Heavyweight Springs INSIDE stanchions		Heavyweight Springs OUTSIDE stanchions	
Length, Max. -	$17^1/_4$ to $17^1/_2$	$17^3/_4$.		$8^1/_2$.	
Length, Min. -	$16^3/_4$	$17^1/_4$.		?	
Wire guage -	9.	8.		0.195.	
Turns/Hand -	82, RH	60, LH or RH		15.	
Compressibility -	?	?		4.285 Kg/cm.	
Outer diameter -	0.745 - 0.755.	0.979 - 0.989.		$1^5/_{16}$.	
Colour Code -	None.	Yellow.	Red/yellow.	Yellow/purple.	Black/yellow.
Part No. -	H848.	H1127	H1434. (Softer spring than H1127).	H1632.	H1865.
Application -	T20, T20C, T20B T20CA, T20J	T20S, T20T 1962 T20SS 1962 T20SR.	1962 TR20 & TS20 All 1963 Sports	TR20 & TS20 1963/4.	All Sports 1965 on

Fork bushes (heavyweight fork) for 1965 on:

	Upper bush	Lower bush
Length -	1.0000.	0.8700 to 0.8750.
Outer diameter -	1.4980 to 1.4990.	1.4935 to 1.4945.
Inner diameter -	1.3065 to 1.3075.	1.2485 to 1.2495.
Tube diameter -	1.3025 to 1.3030.	
Clearance at upper bush -	0.0035 to 0.0050.	
Clearance at lower bush -		0.0035 to 0.0065.
Bore diameter at top of fork -	1.4980 to 1.5000.	

Frame:

Frame dimensions:
  Plunger frames, T15 and T20 - see Fig. A.
  Swinging arm frames, T20 and T20C - see Fig. B.

Front frame geometry:
  Triumph framed machines - Trail = $2^1/_4$.
            Rake = $28^1/_4$.
            Head angle = 65° from the horizontal. = 25° from the vertical.
            Angle between front down tube and steering head = 51°.

Engine location (1960):
  The centre of the conrod was $1^9/_{32}$ to the left of the centre of the frame down tube.
  The gearbox sprocket centre was $2^{13}/_{32}$ to the left of the centre of the frame down tube.

Frame tube sizes and gauges, (swinging fork frames):
  Front down tube $1^1/_8$ OD x 14swg, (0.080).
  Top tube, front frame $1^1/_8$ OD x 14swg, (0.080).
  Seat tube $1^1/_4$ OD x 14swg, (0.080).
  Top tube, rear frame $7/_8$ OD x 16swg, (0.064).
  Down tubes, rear frame $7/_8$ OD x 16swg, (0.064).
  Cross tube, rear frame $3/_4$ OD x 16swg, (0.064).

Petrol tank fixing hole centres = 20in.

Centre stands - (Not Bantam or Super Cub):

  Total leg length, top to bottom:

Wheel size	16	17	18	19
Plunger frames	$6^7/_8$	-	-	$8^5/_8$
Swinging arm frames	$6^7/_8$	$7^3/_4$	$9^7/_8$*	-

                  *=sports type

Steering races:
  Cup and Cone bearings having a 1.2056 pitch circle, with 30 x $1/_4$ balls.

Middle lug and stem (heavyweight forks):
  The angle between the stanchions and the stem should be 0° 40'.

Rear suspension:

Types:
  Plunger type, main spring length = $4^7/_8$.
  Plunger type, rebound spring = $2^3/_4$.

  Swinging fork type -
    Vandervell VP3 bushes working on a $7/_8$ diameter hardened steel pin.
    Bushes were line bored after fitting to 0.8745 to 0.8750.
    End float maximum 0.015. (Spacing washer, and shims of 0.003 and 0.005 used. These were fitted on the right hand side of the swinging fork).

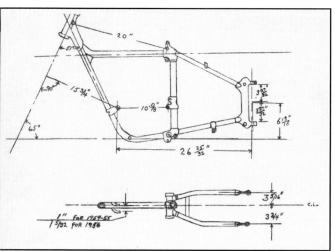

Fig A

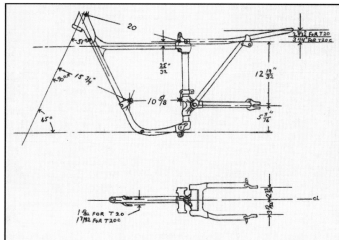

Fig B

Bushes:

   Distance between the suspension unit bush centres -
   'Short' type, for 16 wheels - $10^3/_4$.
   'Long' type, for 17 wheels - $11^1/_4$.
   Adjustable type - $11^1/_4$.

Brakes and wheels:

Brakes:

   Diameter - $5^1/_2$, single leading shoe type, front and rear.

Wheel bearings:

   Front, left side -   6202 type - 15mm ID x 35mm OD x 11mm wide.
   Front, right side - 6203 type - 17mm ID x 40mm OD x 12mm wide.

   Rear, both sides - 6203 type.
   (Sealed wheel bearings were used on the TR20, TS20 and all military models).

Wheel spindle diameter:

   Front, left side -   0.5899 to 0.5904.
   Front, right side - 0.6687 to 0.6692.
   Rear, both sides - 0.6687 to 0.6692.

Wheel offset measurement:

	WM1.19	WM2.16/17
Rear - from the outside edge of rear sprocket to the outside edge of the rim -	$1^1/_8$.	1.
Front - from the outside edge of the brake drum to the edge of the rim -	$^5/_8$.	$^3/_8$.
Rear - from the centre of the rear wheel sprocket to the centre of the rim - any rim	$2^{13}/_{32}$.	

Thread sizes:

   ALL threads were right handed EXCEPT the locking ring for the offside front wheel bearing.
   (See the author's 'Consolidated Parts List and Index' for a complete list of all threaded parts).

Whit/Cycle/Unified	13 tpi:	$^1/_2$.
Whit/Cycle/Unified	14 tpi:	$^5/_8$, $^{11}/_{16}$.
Whit/Cycle/Unified	16 tpi:	$^1/_2$, 1.
Whit/Cycle/Unified	18 tpi:	$^5/_{16}$.
Whit/Cycle/Unified	19 tpi:	$^7/_{16}$, $^1/_2$ (=$^1/_4$ Gas).
Whit/Cycle/Unified	20 tpi:	$^1/_4$, $^3/_8$, $^7/_{16}$, $^1/_2$, $^5/_8$, $^3/_4$, $^7/_8$, 1, $1^1/_8$, $1^1/_4$, $1^1/_2$.
Whit/Cycle/Unified	22 tpi:	$^1/_4$, $^5/_{16}$, $1^1/_{16}$, $1^7/_{16}$, $1^1/_2$.
Whit/Cycle/Unified	24 tpi:	$^{15}/_{16}$.
Whit/Cycle/Unified	26 tpi:	$^1/_8$, $^7/_{32}$, $^1/_4$, $^5/_{16}$, $^3/_8$, $^7/_{16}$, $^1/_2$, $^9/_{16}$, $^{25}/_{32}$, $1^1/_{16}$, $1^5/_8$, $1^5/_8$ (LH), $1^{21}/_{32}$.
Whit/Cycle/Unified	28 tpi:	$^1/_4$, $^3/_8$.
Whit/Cycle/Unified	32 tpi:	$^3/_{16}$, $^{49}/_{64}$, $^{25}/_{32}$, $1^1/_4$.
Whit/Cycle/Unified	40 tpi:	$^7/_{32}$.
BA:		0, 1, 2, 3, 4, 5.
Metric:		12mm, 14mm, 41mm.
Standard Wire Guage:		10, 12, 16 (swg).

Torque settings:

	Ft lb	M Kg
Crankcase bolts -	15	2.1
Cylinder head nuts, 'square' head -	20	2.8
Cylinder head nuts, 'oval' head -	17	2.4
Rocker box nuts -	5	0.7
Rocker feed domed nuts -	9	1.2
Oil pump bolts -	3	0.4
Clutch centre nut -	30*	4.1
Rotor nut -	30*	4.1
Stator nuts -	5	0.7
Front fork pinch bolts -	10	1.4
Fork cap nuts -	15	2.1
Brake cam nuts -	20	2.8

   *Another 'official' source gives a figure of 60ft lb.

Appendix 4

Carburettor - types and settings

The table below has been compiled from all known sources of carburettor information.

		Amal - Zenith -	Choke Size Choke Size	Main Jet Main Jet	Pilot Jet Slow Run Jet	Needle Jet Starter Slide	Needle Pos'n	Needle Type	Slide	Notes
T15	All	Amal 332	$^{11}/_{16}$in	90	20	0.086	3		4	332 /1, /2
T20	3001 - 26275	Amal 332	$^3/_4$in	100	20	0.086	3		4	332 /1, /2, /3, /4, /5.
	26276 - 56359	Zenith 17MX	17mm	78	50	200/65				CS5,C6,C9,C11.
	56360 - 83191	Zenith 18MXZ	18mm	84	45	200/65				C17.
	81890 - end	Amal 32/3	$^{11}/_{16}$in	110	15	0.103	2		2	Export and USA markets.
	83192 - end	Amal 32/1	$^{11}/_{16}$iin	85	15	0.103	2		2	Home market.
T20C	26276 - 35846	Amal 332	$^3/_4$in	100	20	0.086	3		4	332 /3.
	35847 - end	Zenith 17MX	17mm	78	50	200/65				C6,C9,C11.
T20J	36365 - 47347	Amal 332/5 or	$^3/_4$in*	100	20	0.086	3		4	*With restrictor plate 10mm.
	36365 - 47347	Zenith 17MX	17mm*	78	50	200/65				*With restrictor plate 9.5mm.
	47348 - end	Zenith MXZ	9.5 mm							Internal jetting not known.
T20CA	39773 - 39166	Amal 332	$^3/_4$in	100	20	0.086	3		4	
	39167 - end	Zenith 17MXZ	17mm	78	50	200/65				CS5.
T20S	Scrambler	Amal 376/217	$^{15}/_{16}$in	140	20	0.106	3	C	3	Remote float.
	Racing	Amal 376/217	$^{15}/_{16}$in	200	20	0.106	4	C	3	Remote float.
	Road	Amal 376/272	$^{15}/_{16}$in	140	20	0.106	3	C	3	or 18mm Zenith.
	Trials	Zenith 18MX	18mm*	84	45	200/65				*20mm Zenith also used.
	Trials, (TriCor)	Amal 332	$^7/_8$in	140	15	0.086	3		4	332/7. 'Woods' model.
T20T	All	Zenith 18MX	18mm	84	45	200/65				
	alternative	Amal 375/44	$^{25}/_{32}$in	100	25	0.105	3	B	$3^1/_2$	Military machines.
T20W	All	Zenith 18MX	18mm	84	45	200/65				Internal jetting assumed.
T20SL	All	Amal 376/272	$^{15}/_{16}$in	140	20	0.106	3	C	3	
T20SS	USA east coast	Amal 376/272	$^{15}/_{16}$in	140	20	0.106	3	C	$2^1/_2$	All other markets.
	USA west coast	Amal 376/272	$^{15}/_{16}$in	160	25	0.106	3	C	$2^1/_2$	
T20SH	All	Amal 376/272	$^{15}/_{16}$in	140	20	0.106	3	C	3	
T20SC	All	Amal 376/272	$^{15}/_{16}$in	140	20	0.106	3	C	3	
T20SR	All, except below	Amal 376/272	$^{15}/_{16}$in	140	20	0.106	3	C	3	
	USA west 1965/6	Amal 376/314	$^{15}/_{16}$in	140	15	0.106	3	C	$2^1/_2$	
TR20	All	Amal 375/44	$^{25}/_{32}$in	100	25	0.105	3	B	$3^1/_2$	
TS20	All, except below	Amal 376/272	$^{15}/_{16}$in	140	20	0.106	3	C	3 or $2^1/_2$	
	USA west coast	Amal 376/272	$^{15}/_{16}$in	180	20	0.106	3	C	3	
T20WD	All	Amal 375/44	$^{25}/_{32}$in	100	25	0.105	3	B	$3^1/_2$	
T20SM	All	Amal 376/314	$^{15}/_{16}$in	140	15	0.106	3	C	$3^1/_2$*	*Some sources show $2^1/_2$.
T20M	All	As T20SM.								
T20MWD	French Army	Amal 375/44	$^{25}/_{32}$in	100	25	0.105	3	B	$3^1/_2$	
T20P	All	Amal 32/1	$^{11}/_{16}$in	85	15	0.103	2		2	Type & jetting assumed.
T20B	Bantam Cub	Amal 375/61	$^{25}/_{32}$in	90	25	0.105T	3	B	$3^1/_2$	
	Super Cub to 1967	Amal 375/61	$^{25}/_{32}$in	90	25	0.105T	3	B	$3^1/_2$	
	Super Cub 1968	Amal R622/1	22mm	120	25	0.105	2		3	Concentric.

Appendix 5

Model colour schemes

Listed below are the specified standard colour schemes for each year and model taken from various sales brochures and other literature. These sources often listed models that did not sell in that year, so those models have not been included.

The sales brochures were usually created many months before the model went into production and, by the time it did, the colours or the names of the colours had sometimes been changed.

1954	All markets	T15	Amaranth Red.
		T20	Shell Blue Sheen and black.
1955	UK & General Export	T15	Amaranth Red.
		T20	Shell Blue Sheen and black.
	USA	T15	Black.
		T20	Shell Blue Sheen and black.
1956	All Markets	T15	Amaranth Red.
		T20	Shell Blue Sheen and black.
1957	All markets	T20	Crystal Grey and black.
		T20C	As T20.
1958	UK & General Export	T20	Crystal Grey and black.
		T20C	As T20.
	USA	T20	Aztec Red and black.
		T20C	As T20.
		T20J	As T20.
		T20CA	As T20.
1959	UK & General Export	T20	Crystal Grey and black.
		T20C	As T20.
		T20S	Ivory, Azure Blue and black.
	USA	T20	Aztec Red and black.
		T20C	As T20.
		T20J	As T20.
		T20CA	As T20.
		T20S	Ivory, Azure Blue and black, or Aztec Red, Ivory and black.
1960	UK & General Export	T20	Crystal Grey and black.
		T20S	Ivory, Azure Blue and black.
	USA	T20	Aztec Red and black.
		T20J	As T20.
		T20S	Ivory, Azure Blue and black.
		T20W	As T20S.
1961	UK & General Export	T20	Silver Sheen and black.
		T20S	Ruby Red[1], Silver Sheen and black.
		T20T	As T20S.
		T20SL	As T20S.
		T20WD	Deep Bronze Green (BSC 224) - (Military T20T machines).
	USA	T20	Silver Sheen and black.
		T20J	as T20.
		T20S	Ruby Red[1], Silver Sheen and black.
		T20T	As T20S.
		T20SL	As T20S.
1962	UK	T20	Silver Sheen and black.
		T20SS	Burgundy[2], Silver Sheen and black.
		T20SH	As T20SS.
		TR20	As T20SS.
		TS20	As T20SS.
	USA & General Export	T20	Flame[3], Silver Sheen and black.
		T20SS	Burgundy[2], Silver Sheen and black.
		T20SC	As T20SS.
		T20SR	As T20SS.
		TR20	As T20SS.

1963	UK & General Export	T20	Flame[3], Silver Sheen & black.
		T20SS	Flamboyant Flame[3], Silver Sheen and black.
		T20SH	Burgundy[2], Silver Sheen & black.
		TR20	As T20SH.
		TS20	As T20SH.
		T20WD	NATO Green, (Olive Drab, BSC 298).
	USA	T20	Flamboyant Ruby Red[2], Silver Sheen & black.
		T20SS	Flamboyant Flame[3], Silver Sheen & black.
		T20SC	As T20SS.
		T20SR	As T20SS.
		TR20	No information, probably as T20 or T20SS.
		TS20	As TR20.
1964	UK & General Export	T20	Hi Fi Scarlet[2], Silver Sheen & black.
		T20SS	Crystal Blue[4], Silver Sheen and black.
		T20SH	As T20.
		TR20	As T20.
		T20WD	NATO Green, (Olive Drab, BSC 298).
	USA	T20	Flamboyant Scarlet[2], Silver Sheen & black.
		T20SS	Crystal Blue [4], Silver Sheen & black.
		T20SC	Kingfisher Blue [4], Silver Sheen & black.
		T20SR	Kingfisher Blue [4], Silver Sheen & black.
		T20SM	Crystal Blue[4], Silver Sheen and black.
1965	UK & General Export	T20	Hi Fi Scarlet[2], Silver Sheen & black.
		T20SS	Kingfisher Blue[4], Silver Sheen and black.
		T20SH	As T20.
		TR20	As T20.
	USA	T20	Flamboyant Scarlet[2], Silver Sheen & black.
		T20SC	Hunting Yellow & black.
		T20SR	Pacific Blue[4], Silver Sheen & black.
		T20SM	As T20SC.
1966	UK & General Export	T20	Hi Fi Scarlet[2], Silver Sheen & black.
		T20SH	As T20. Later, Metallic Blue[5], Alaskan White & black.
		T20B	Bantam Cub. Nutley Blue[6], Alaskan White & black.
	USA	T20	Flamboyant Scarlet[2], Silver Sheen & black.
		T20M	Grenadier Red[7], Alaskan White & black.
1967	UK & General Export	T20B	Bantam Cub. Nutley Blue[6], Alaskan White & black.
		T20B	Super Cub. Bushfire Red[8], black & chrome.
		T20SH	Metallic Blue[5], Alaskan White & black.
		T20M	Grenadier Red[7], Alaskan White & black.
		T20MWD	NATO Green, (Olive Drab, BSC 298).
	USA	T20M	Grenadier Red[7], Alaskan White & black.
1968	UK & General Export	T20B	Bantam Cub. Nutley Blue[6], Alaskan White & black
		T20B	Super Cub. Firecracker Red[8], black & chrome.
		T20B	'Tarbuk'. Grenadier Red?[7], Silver Sheen? & black.
		T20MWD	NATO Green, (Olive Green BSC 298).
	USA	T20M	Grenadier Red[7], Alaskan White & black.

Notes:

1. Ruby Red was a solid colour also known in the USA as Bright Amaranth Red.
2. Various translucent reds - Burgundy, Flamboyant Ruby Red, Flamboyant Red, Hi Fi Scarlet, Flamboyant Scarlet and Sapphire Red - may all have been different names for the same colour. Applied over gold basecoat for 1962/3 and a silver basecoat from 1964 onwards.
3. Flame and Flamboyant Flame are thought to be the same translucent colour. Applied over a gold basecoat.
4. Kingfisher Blue and Pacific Blue are thought to be different names for the same translucent colour. Crystal Blue was a slightly lighter shade. Applied over a silver basecoat.
5. Metallic Blue was a translucent colour applied over a silver basecoat.
6. Nutley Blue was a solid colour also shown as Perrivale Blue in some literature. Other literature shows this model in Pacific Blue (the same as Kingfisher Blue) but as far as is known no machines were built in this colour.
7. Grenadier Red was a solid colour applied over a white basecoat.
8. Bushfire Red and Firecracker Red are thought to be different names for the same colour.

Appendix 6

Delivery details - by country of delivery

Notes to Appendix 6.

The figures in this appendix have been arrived at from close examination and summary of the existing factory despatch and build books. These records are known to contain inaccuracies and a degree of interpretation has sometimes been applied to try and make some allowances for these shortcomings.

This philosophy not only applies to the model type but also the country of destination. The country names are those shown in the factory records, plus the modern name. This detail was sometimes not evident in the records and in these instances it has had to be deduced from other information, such as the customer's initials or the name of a town. These machines are shown as 'Export' in the figures for the country to which they are considered most likely to have been sent. If no reasonable guess can be made then they are shown in the 'Unknown' heading and in a likely geographical region. The figures shown in this Appendix must not therefore be taken as definitive totals, although they are the best estimates that can be made in the light of present knowledge.

The penultimate column, 'BSA', requires some explanation. At Meriden, the opening and closing machine serial numbers for each year were known, having been taken from one sequence of numbers - 101 to 100013 - that was used for all models.

The BSA group used a different numbering method whereby several series of numbers, starting at no.101, no.2001 or no.3001, were used for the various Tiger Cub models. The opening and closing serial numbers for each season have not been recorded and it is therefore impossible to say to what 'year' any particular machine belongs. For this reason the figure shown in the BSA column includes all the machines built at Small Heath from 1965 to 1969, plus the small number of 1965 machines delivered from Meriden with serial numbers between 99720 and 100013, plus any 1965 Meriden built machines with numbers after the fresh start at no.101.

Country	Model	1954	1955	1956	1957	1958	1959	1960	1961	1962	1963	1964	BSA	Totals
U.K. - England	T15	2471	1358	305										4134
	T20	924	3831	3948	3920	4194	5256	6643	5639	1998	1873	1422	1359	41007
	ISDT			1										1
	T20C				506	608	272							1386
	T20J						1							1
	T20S						134	655	1				1	791
	T20T						1	2	253	1			2	259
	T20SL							1	1153					1154
	T20SS									13	1		3	17
	T20SH									531	314	203	532	1580
	TR20									186	55	79	87	407
	TS20									10	2			12
	T20WD										4			4
	T20SM												45	45
	T20P												12	12
	B.Cub												847	847
	T20M												116	116
	S.Cub												1274	1274
	Tarbuk												46	46
	'Export'												9	9
	All	3395	5189	4254	4426	4802	5664	7301	7046	2739	2249	1704	4333	53102
	Engines	4	1				1		13	27	6	21	8	81
U.K. - Isle of Man	T15	5	6	1										12
	T20	2	8	8	12	9	4	11	8	2	6			70
	T20C				6	7	2							15
	T20S							2						2
	T20SL								5					5
	T20SH										2			2
	TR20									2	1	1	1	5
	B.Cub												1	1
	S.Cub												1	1
	All	7	14	9	18	16	6	13	13	4	9	1	3	113
U.K. - N. Ireland	T15	51	21	7										79
	T20	22	78	98	165	202	225	200	135	64	56	36	19	1300
	T20C				13	35	14							62
	T20S						4	13						17
	T20SL								18					18
	T20SH										2	1	4	7
	TR20									3			1	4
	TS20									2				2
	B.Cub												6	6
	S.Cub												17	17
	All	73	99	105	178	237	243	213	153	69	58	37	47	1512
U.K. - Scotland	T15	115	59	32										206
	T20	45	236	329	403	466	539	596	494	172	147	90	67	3584
	T20C				30	27	9							66
	T20S						13	37						50
	T20T								19					19
	T20SL								81					81
	T20SH									28	14	3	18	63
	TR20									2	1			3
	TS20									3				3
	B.Cub												49	49
	T20M												1	1
	S.Cub												61	61
	All	160	295	361	433	493	561	633	594	205	161	94	196	4186

Country	Model	1954	1955	1956	1957	1958	1959	1960	1961	1962	1963	1964	BSA	Totals
U.K. - Wales	T15	130	57	23										210
	T20	57	183	218	226	207	202	277	223	88	89	58	53	1881
	T20C				15	20	5							40
	T20S						3	13						16
	T20T								6					6
	T20SL								28					28
	T20SH									24	7	3	13	47
	TR20									7		4	5	16
	TS20									1				1
	B.Cub												36	36
	T20M												1	1
	S.Cub												77	77
	All	187	240	241	241	227	210	290	257	120	96	65	185	2359
U.K. - Region not known	T20		25	2	1	2	3				3	8	92	136
	T20SL								1					1
	T20SH										1		8	9
	TR20												4	4
	T20P												4	4
	S.Cub												2	2
	All		25	2	1	2	3		1		4	8	110	156
U.K. - All regions	T15	2772	1501	368										4641
	T20	1050	4361	4603	4727	5080	6229	7727	6499	2324	2174	1614	1590	47978
	ISDT				1									1
	T20C				570	697	302							1569
	T20J						1							1
	T20S						154	720	1				1	876
	T20T						1	2	278	1			2	284
	T20SL						1	1286						1287
	T20SS									13	1		3	17
	T20SH									583	340	210	575	1708
	TR20									200	56	85	98	439
	TS20									16	2			18
	T20WD										4			4
	T20SM												45	45
	T20P												16	16
	B.Cub												939	939
	T20M												118	118
	S.Cub												1432	1432
	Tarbuk												46	46
	'Export'												9	9
	All	3822	5862	4972	5297	5777	6687	8450	8064	3137	2577	1909	4874	61428
	Engines	4	1					1	13	27	6	21	8	81
Aden/South Yemen	T15	4		2										6
	T20		2	4	8	12	12	12	14	13	23	4	4	108
	T20C					1								1
	T20S								1					1
	T20SL									2				2
	T20SS										1			1
	B.Cub												1	1
	S.Cub												2	2
	All	4	2	6	8	13	12	13	16	14	23	4	7	122
Algeria	T15	9												9
	T20		8	6	19									33
	T20S						2							2
	All	9	8	6	19		2							44
Antigua	T15	8	7											15
	T20		3	6		2	3	2			1			17
	All	8	10	6		2	3	2			1			32
Argentina	T20 only							8	48					56
Ascension Is.	T20							1	1				2	4
	T20S							3						3
	T20SS										1			1
	S.Cub												1	1
	All							4	1			1	3	9
Australia	T15	456	56	18										530
	T20	44	181	163	133	84	50	28	8	5	10		1	707
	T20C				4	3	4							11
	T20S						1	21	6					28
	T20T								16					16
	T20SL								7					7
	T20SS										7			7
	T20SM												1	1

181

Country	Model	1954	1955	1956	1957	1958	1959	1960	1961	1962	1963	1964	BSA	Totals
Australia (cont'd)	T20M												3	3
	'Export'												3	3
	All	500	237	181	137	87	55	49	37	5	17		8	1313
Bahamas	T15	3												3
	T20	1	2		2		1	2	3	2		4	2	19
	B.Cub												1	1
	'Export'												1	1
	All	4	2		2		1	2	3	2		4	4	24
Bahrain	T15	4												4
	T20		2				1			2	2			7
	T20SL								2					2
	All	4	2				1		2	2	2			13
Barbados	T15	15												15
	T20	3	8	8	14	22	16	18	24	8	6	8	4	139
	B.Cub												4	4
	S.Cub												4	4
	All	18	8	8	14	22	16	18	24	8	6	8	12	162
Belgium	T15	56												56
	T20	18	41	10	12	47	42	147	99	70		5	1	492
	T20C				4	2								6
	T20S						10	6						16
	T20T								13					13
	T20SS									15				15
	S.Cub												1	1
	All	74	41	10	16	49	52	153	112	85		5	2	599
Bermuda	T15	121	99	90										310
	T20			47	61	136	146	140	88	60	10	30		718
	T20C				1	2	2							5
	T20J								2					2
	T20SS									2	10	2		14
	All	121	99	137	62	138	148	140	90	62	20	32		1049
Bolivia	B.Cub only											1		1
British Guiana/Guyana	T15	4	4	16										24
	T20		16	12	6	7	4	12	10	12		18	5	102
	T20C				2	11	6							19
	T20S						2	4						6
	T20SH												1	1
	T20WD										3			3
	B.Cub												6	6
	S.Cub												13	13
	All	4	20	28	8	18	12	16	10	12		21	25	174
British Honduras/Belize	T15 only	6	2											8
British North Borneo/Sabah	T15	4												4
	T20	2	4	22	14	14	13	8	11	10	8	6	2	114
	T20C						2							2
	T20SL								1					1
	B.Cub												2	2
	All	6	4	22	14	14	15	8	12	10	8	6	4	123
Brunei	T15	7												7
	T20			2	4	4	6	10	14	8	4	6		58
	T20SM												2	2
	All	7		2	4	4	6	10	14	8	4	6	2	67
Bulgaria	T20 only						2							2
Burma	T15	2												2
	T20		2						225					227
	All	2	2						225					229
Cambodia	T20 only				2			3	2					7
Cameroon Republic	T15	4												4
	T20			4	15									19
	All	4		4	15									23
Canada	T15	75	4											79
	T20	18	38	31	39	27	25	22	11	5	7	7	14	244
	T20C				11	21	5							37
	T20CA					1								1
	T20S						23	16	2					41

Country	Model	1954	1955	1956	1957	1958	1959	1960	1961	1962	1963	1964	BSA	Totals
Canada (cont'd)	T20T								3					3
	T20SL								9					9
	T20SS									16	4	1	3	24
	T20SH									5	5	1	9	20
	T20SC											6		6
	T20SR												4	4
	TS20												2	2
	T20SM											1	75	76
	B.Cub												16	16
	T20M												222	222
	S.Cub												13	13
	'Export'												16	16
	All	93	42	31	50	49	53	38	25	26	16	16	374	813
Canary Is.	T15	15	6	1										22
	T20		3			1					9	11	3	27
	B.Cub												3	3
	All	15	9	1		1					9	11	6	52
Ceylon/Sri Lanka	T15	24	16	19										59
	T20	4	45	147	151	54	136	149	46					732
	T20C				11	5	16							32
	T20CA					1								1
	T20S						11							11
	B.Cub												4	4
	T20M												11	11
	S.Cub												142	142
	'Export'												1	1
	All	28	61	166	162	60	152	160	46				158	993
Channel Is.	T15	19	12	3										34
	T20	7	23	47	32	45	54	53	45	12	12	7	21	358
	T20C				1	1	1							3
	T20S						1	8						9
	T20T								5					5
	T20SL								11					11
	T20SH									7	1	7	14	29
	TR20									1		1		2
	T20SM												4	4
	B.Cub												8	8
	S.Cub												15	15
	All	26	35	50	33	46	56	61	61	20	13	15	62	478
Chile	T15	1		4										5
	T20		4	3	3	1	2	20	7					40
	B.Cub												1	1
	All	1	4	7	3	1	2	20	7				1	46
Colombia	T15	49	12											61
	T20	7	27											34
	All	56	39											95
Congo	T20					4	10							14
	T20S							4						4
	All					4	10	4						18
Costa Rica	T20			2	15	26	24	30						97
	T20C						1							1
	T20S						1							1
	All			2	15	26	26	30						99
Cuba	T15	26	4											30
	T20	1	5		2									8
	All	27	9		2									38
Cyprus	T15	6	4	2										12
	T20	5	12	24	18	21		4	1	4	6			95
	T20C				1	1								2
	All	11	16	26	19	22		4	1	4	6			109
Denmark	T15	34												34
	T20	17	123	375	44		20					1		580
	T20C				35									35
	T20S						5	4	3					12
	T20SL								3					3
	All	51	123	375	79		25	4	6			1		664
Dominica	T15	4		3										7
	T20	2	3	4	2			2	2	8	6		2	31
	All	6	3	7	2			2	2	8	6		2	38

Country	Model	1954	1955	1956	1957	1958	1959	1960	1961	1962	1963	1964	BSA	Totals
Dubai	T20 only											2		2
Dutch Guiana/Surinam	T20 only						1	1						2
Ecuador	T15		2											2
	T20		4	9	4	1		2				1	5	26
	T20SS												1	1
	T20SH												3	3
	All		6	9	4	1		2				1	9	32
Eire	T15	110	20	18										148
	T20	21	105	77	69	67	95	90	122	146	135	112	95	1134
	T20C				1		1							2
	T20S							1						1
	T20SL								2					2
	T20SS									3				3
	T20SH												1	1
	TR20									1		3		4
	B.Cub												60	60
	T20M												1	1
	S.Cub												104	104
	All	131	125	95	70	68	95	91	124	150	135	115	261	1460
Ethiopia	T15 only	4												4
Falkland Is.	T20					3	2		1		1			7
	T20C				1	3	4							8
	T20S						1	2	1					4
	T20T								3					3
	T20SL								1					1
	T20SS									1	1			2
	All				1	6	7	2	6	1	2			25
	Engines										1			1
Faroe Is.	T20 only			1	1									2
Fernando Póo/Bioko	T15 only	2												2
Fiji	T15	4	4	1										9
	T20		5	6	1	1	1		3			1	2	20
	B.Cub												1	1
	All	4	9	7	1	1	1		3			1	3	30
Finland	T15			9										9
	T20			18	33	88	30	37	107	105	54			472
	T20C						6							6
	T20CA						1							1
	T20S						3	1						4
	T20T								1					1
	T20SL								13					13
	T20SS									13	20			33
	TR20									2	5	6		13
	TS20									1	1			2
	All			27	33	88	40	38	121	121	80	6		554
	Engines								12		1			13
France	T15	3		4										7
	T20		1	28	106	7	4	12	4	3	3	10	14	192
	T20C					1	1							2
	T20S						1	2						3
	T20T								1					1
	T20SS									12				12
	T20SH											1		1
	TR20										4	3	1	8
	TS20									3	4	1		8
	T20WD											81	2	83
	T20SM												722	722
	B.Cub												3	3
	T20M												478	478
	S.Cub												116	116
	'Export'												18	18
	All	3	1	32	106	8	6	14	5	18	11	96	1354	1654
	Engines							1						1
French Somaliland/Djibouti	T15 only	4	1											5
Friendly Is./Tonga	T20 only			6	5									11
Gambia	T20C only						2							2

Country	Model	1954	1955	1956	1957	1958	1959	1960	1961	1962	1963	1964	BSA	Totals
Germany	T15	18												18
	T20	2		2	4	6	1	2		1				18
	T20C				1									1
	T20S						1	1						2
	T20SS									1				1
	T20SH												1	1
	S.Cub												3	3
	All	20		2	5	6	2	3		2			4	44
Gibraltar	T15	1												1
	T20	1	6	4	6	1	3	1	2		1			25
	S.Cub												1	1
	All	2	6	4	6	1	3	1	2		1		1	27
Gold Coast/Ghana	T20 only						3		5			2		10
Greece	T15	9	3											12
	T20					1								1
	S.Cub												1	1
	All	9	3			1							1	14
Grenada	T15	4												4
	T20		1					6	8	4	2	2	2	25
	B.Cub												3	3
	All	4	1					6	8	4	2	2	5	32
Guadeloupe (French Antilles)	T15	2	1	1										4
	T20		1		1									2
	All	2	2	1	1									6
Guam	T20		16	12		2		1	8		2		6	47
	T20C				8	2	4							14
	T20S						2	7						9
	T20T								1					1
	T20SL								3					3
	T20SS									4	2			6
	T20SH										2		5	7
	T20SR											4		4
	TR20									4				4
	TS20									4				4
	T20SM											4	33	37
	T20M												6	6
	'Export'												2	2
	All		16	12	8	4	6	8	12	12	6	8	52	144
Guatemala	T15	1												1
	T20		4	8	24	46	114	10	50	8	20			284
	T20C					2								2
	T20S							2						2
	B.Cub												1	1
	All	1	4	8	24	48	114	12	50	8	20		1	290
Haiti	T15	2												2
	T20			2	2				1					5
	All	2		2	2				1					7
Holland	T15	89	10	2										101
	T20	5	19	3		2	7		2		4			42
	T20T								1					1
	T20SL								1					1
	TR20									1				1
	All	94	29	5		2	7		4	1	4			146
Hong Kong	T15	4	3											7
	T20	1	13	22	31	22	14	16	8	16	29	120	98	390
	T20C				1	1								2
	T20SS										8			8
	T20SH												4	4
	T20SR											1		1
	T20WD											1		1
	B.Cub												79	79
	S.Cub												29	29
	All	5	16	22	32	23	14	16	8	16	37	122	210	521
India	T15	86	42	21										149
	T20	19	41	104	67	1		2				1	3	238
	T20SM												1	1
	All	105	83	125	67	1		2				1	4	388

Country	Model	1954	1955	1956	1957	1958	1959	1960	1961	1962	1963	1964	BSA	Totals
Indo China/South Vietnam	T15			5										5
	T20			15										15
	T20SM												1	1
	T20M												3	3
	All			20									4	24
Indonesia	T15	2		124	30									156
	T20			76	50				101					227
	All	2		200	80				101					383
Iran	T15	6	3	4										13
	T20			19	5	32	195				20	39	89	399
	T20C					19								19
	T20CA					1								1
	B.Cub												139	139
	S.Cub												142	142
	All	6	3	23	5	52	195				20	39	370	713
	Engines												3	3
Iraq	T20 only				1							2	3	6
Israel	T20 only								1			4	4	9
Italy	T15	12												12
	T20	6	7		13	5	3	1			3			38
	T20C				9									9
	T20S						2	4						6
	T20SS										1			1
	TR20											1		1
	All	18	7		22	5	5	5			4	1		67
Ivory Coast	T20C only						1							1
Jamaica	T15	23	4											27
	T20	6	6	11	24	10	12	12	7	8	12	18	22	148
	T20SS											6		6
	T20SR											2		2
	T20WD											1		1
	T20SM												11	11
	B.Cub												38	38
	T20M												15	15
	S.Cub												203	203
	'Export'												4	4
	All	29	10	11	24	10	12	12	7	8	12	27	293	455
Japan	T15	10	1											11
	T20		49	15	15	4	4	2						89
	T20C				7	4	3							14
	T20S						3	3						6
	T20SL								2					2
	T20SH									1				1
	TR20										1			1
	TS20										1			1
	T20SM												1	1
	All	10	50	15	22	8	10	5	2	1	2		1	126
Java	T20 only		1											1
Jordan	T20 only		2	3	3	2								10
Kenya	T15	15	2											17
	T20	3	4	6	22	10	18	15	25	12	44	60	103	322
	T20S								2					2
	T20SL								5					5
	T20SM												24	24
	B.Cub												34	34
	T20M												17	17
	'Export'												33	33
	All	18	6	6	22	10	18	15	32	12	44	60	211	454
Kuwait	T15 only		10											10
Laos	T20												22	22
	T20SS												6	6
	All												28	28
Lebanon	T15 only	1												1
Liberia	T20 only					4								4

Country	Model	1954	1955	1956	1957	1958	1959	1960	1961	1962	1963	1964	BSA	Totals
Libya	T15	1												1
	T20				2					7				9
	All	1			2					7				10
Madeira	T15 only	1												1
Malaya	T15	40												40
	T20	7	37	93	92	106	112	218	375	369	260	78	12	1759
	T20C					2								2
	T20S							1						1
	T20SL									2				2
	T20SS										1	2		3
	All	47	37	93	92	108	112	219	377	369	261	80	12	1807
Malaysia	B.Cub only												12	12
Maldive Is.	T20 only							1						1
Malta	T15	5												5
	T20	2	5	9	8	26	36	15	20	4	2		3	130
	T20SL								1					1
	T20SS										4	2		6
	S.Cub												8	8
	All	7	5	9	8	26	36	15	21	4	6	2	11	150
Martinique	T20 only					2		4	2		2	2		12
Mauritius	T15	3	2											5
	T20				2			2			2			6
	All	3	2		2			2			2			11
Mexico	T15	26												26
	T20	4	3	12	29	16	19	7		12	2			104
	T20C				23	4								27
	T20CA						3							3
	T20S						1	7	1					9
	T20SS										1			1
	TS20										2			2
	S.Cub												3	3
	All	30	3	12	52	20	23	14	1	12	5		3	175
Montserrat	T20 only				1		1		1		1			4
Morocco	T15	28												28
	T20	1		22	22	3	2	2						52
	T20C				4									4
	S.Cub												1	1
	All	29		22	26	3	2	2					1	85
Nepal	T20				1					1	4	8	12	26
	T20SH										1			1
	B.Cub												10	10
	S.Cub												37	37
	All				1					1	5	8	59	74
New Zealand	T15	173	15											188
	T20	27	57	4	36	31	32	50	48	43	74	24	18	444
	T20C				2	2								4
	T20S							3						3
	T20SS												1	1
	T20SH										6		8	14
	B.Cub												25	25
	S.Cub												5	5
	'Export'												2	2
	All	200	72	4	38	33	32	53	48	43	80	24	59	686
	Engines											1		1
Nicaragua	T15	22												22
	T20	11		8	21	24	24	71	72	58	86	90	36	501
	T20S						6							6
	T20SL								10					10
	T20SS									4	14	6		24
	T20SM												2	2
	B.Cub												34	34
	'Export'												5	5
	All	33		8	21	24	24	77	82	62	100	96	77	604
Niger	T20SS only										1			1
Nigeria	T15		15											15

187

Country	Model	1954	1955	1956	1957	1958	1959	1960	1961	1962	1963	1964	BSA	Totals
Nigeria (cont'd)	T20	3	2	11	28	65	233	291	253		52	29	71	1038
	B.Cub												35	35
	S.Cub												6	6
	All	18	2	11	28	65	233	291	253		52	29	112	1094
North East New Guinea	T15			3										3
	T20			6	3	6		6						21
	All			9	3	6		6						24
Northern Rhodesia/Zambia	T20		9	14	21	4	2	6	4	3	4	2	1	70
	T20C						1							1
	T20S							2						2
	T20SH											1		1
	All		9	14	21	4	3	8	4	3	4	3	1	74
Norway	T15	73	32	72										177
	T20	20	6	27	67	66	31	36	27	13	5	2		300
	T20C				4	3	6							13
	T20S						2	1						3
	T20SL								1					1
	T20SS										1			1
	S.Cub												2	2
	All	93	38	99	71	69	39	37	28	13	6	2	2	497
Nyasaland/Malawi	T15		4											4
	T20				10	8	6	10	8					42
	T20SL								2					2
	T20SS									3				3
	All		4		10	8	6	10	10	3				51
Okinawa	T20			7		1								8
	T20C					3	1							4
	T20SM												2	2
	All			7		4	1						2	14
Oman	T20 only				4	4								8
Pakistan	T15	25	4	8										37
	T20	6	8	28	19	21	22	31	196	583	828	654	99	2495
	T20S						1							1
	T20T											5		5
	T20SS									1				1
	T20SH									1				1
	T20WD										2			2
	T20SM												4	4
	B.Cub												22	22
	S.Cub												2	2
	All	31	12	36	19	21	23	31	196	585	828	656	132	2570
Panama Canal Zone	T15	5												5
	T20	1	2	2										5
	T20SH										1		1	2
	T20SM												1	1
	All	6	2	2							1		2	13
Papua/New Guinea	T20			2		2			2	3				9
	T20T								1					1
	T20SL								2					2
	T20SS										1			1
	All			2		2			5	3	1			13
Paraguay	T15		2											2
	T20		5			11			1					17
	T20C					2								2
	T20SH									4			2	6
	All		7			13			1	4			2	27
Peru	T15	10												10
	T20		2	6	23	22	39	16	5	4	3			120
	T20SM												1	1
	All	10	2	6	23	22	39	16	5	4	3		1	131
Philippines	T15	4												4
	T20				6		3		15		10	20	19	73
	T20SL								3					3
	T20SS											3		3
	S.Cub												5	5
	All	4			6		3		18		10	23	24	88

Country	Model	1954	1955	1956	1957	1958	1959	1960	1961	1962	1963	1964	BSA	Totals
Poland	T20							1	1		1			3
	T20C				1									1
	All				1			1	1		1			4
Portugal	T15	15												15
	T20	4	8	45	50	24	24	46	20	8	20	36	16	301
	T20C						1							1
	T20S						1							1
	B.Cub												24	24
	S.Cub												28	28
	All	19	8	45	50	24	26	46	20	8	20	36	68	370
Portugese E. Africa/Mozambique	T15	16	5											21
	T20	3	8	1	2	2	3	10	26	2	9	16	8	90
	T20SL								2					2
	T20SS									4				4
	B.Cub												4	4
	S.Cub												6	6
	All	19	13	1	2	2	3	10	28	6	9	16	18	127
Portugese Guinea	T20 only				2	3		6	1	2		3	3	20
Portugese India/Goa	T20 only	2												2
Portugese West Africa/Angola	T15	4		4										8
	T20			2	25	57	47	108	63	80	85	66	48	581
	T20SS									3				3
	B.Cub												12	12
	All	4		6	25	57	47	108	63	83	85	66	60	604
Puerto Rico	T15	8												8
	T20							4	12	4	2	7		29
	T20CA					5	6							11
	T20SL									2				2
	T20SS									2				2
	All	8				5	6	4	14	6	2	7		52
Reunion Is.	T15				2									2
	T20		2		4			2						8
	All		2		6			2						10
Rumania	T20 only				1									1
Salvador	T20		1			2					1			4
	T20C					2	1							3
	T20S						1							1
	All		1			4	2				1			8
Sarawak	T15	18	2											20
	T20	2	4	4		1	4	4	5	13	10	10	3	60
	T20C					1								1
	B.Cub												10	10
	All	20	6	4		2	4	4	5	13	10	10	13	91
Saudi Arabia	T20 only												1	1
Sierra Leone	T20				11	9	3	18	36	12	11	14	14	128
	B.Cub												6	6
	All				11	9	3	18	36	12	11	14	20	134
Singapore	T15	12												12
	T20	2	41	148	129	75	157	166	294	190	102	49	38	1391
	T20C					17								17
	T20S						6							6
	T20SL									2				2
	T20SS									1				1
	T20SH									1	1			2
	B.Cub												8	8
	All	14	41	148	129	92	163	166	296	192	103	49	46	1439
Society Is./Tahiti	T15	2												2
	T20				1			2						3
	All	2			1			2						5
South Africa	T15	28	2											30
	T20	5	30	135	146	180	79	71	59	34	38	30	17	824
	T20C					16								16
	T20S						1							1
	T20SS									3	3			6
	T20SH									1		2		3

Country	Model	1954	1955	1956	1957	1958	1959	1960	1961	1962	1963	1964	BSA	Totals
South Africa (cont'd)	TR20										1			1
	TS20										1			1
	B.Cub												2	2
	T20M												2	2
	S.Cub												7	7
	All	33	32	135	146	196	80	71	59	38	43	32	28	893
South Korea	T20 only						3	12	5					20
South West Africa/Namibia	T20 only				7									7
Southern Rhodesia/Zimbabwe	T20		3	11	14	27	5	7	11	18	3			99
	T20C				2	7	1							10
	T20S							2						2
	T20T								2					2
	T20SL								3					3
	T20SS									1				1
	T20SM												1	1
	All		3	11	16	34	6	9	16	19	3		1	118
Spain	T15	1												1
	T20									1	12			13
	T20SH										4			4
	All	1								1	16			18
Spanish Guinea	T15		3											3
	T20		1		1						1			3
	All		4		1						1			6
St. Helena	B.Cub												1	1
	S.Cub												1	1
	All												2	2
St. Kitts	T20								4	4	2	2	6	18
	B.Cub												2	2
	All								4	4	2	2	8	20
St. Lucia	T20 only			4	6		3	2						15
St. Vincent	T15		2											2
	T20		4		8		2						2	16
	B.Cub												2	2
	All		6		8		2						4	20
Sudan	T15	2		1										3
	T20	2	1	2	2			4						11
	T20T								1					1
	All	4	1	3	2			4	1					15
Sweden	T15	150												150
	T20	27	17											44
	T20C					2								2
	T20S						1	4						5
	TR20									1	1	1		3
	All	177	17			2	1	4		1	1	1		204
	Engines									1				
Switzerland	T15	14												14
	T20	10		5	5	1		1		1	4	3		30
	T20S							1						1
	T20SH										2	1		3
	TR20									1		1		2
	All	24		5	5	1		2		2	6	4	1	50
Syria	T15	7												7
	T20	1											1	2
	T20SM												1	1
	All	8											2	10
Taiwan (Formosa)	T20						1	1	6	5				13
	T20SS									3	4	2		9
	T20SH												5	5
	B.Cub												5	5
	All						1	1	6	8	4	2	10	32
Tanganyika/Tanzania	T15	12		3										15
	T20	5	79	32	16	10	25	20	31	47	34	10	14	323
	T20C					2	2							4
	T20CA						2							2
	T20S							2						2

Country	Model	1954	1955	1956	1957	1958	1959	1960	1961	1962	1963	1964	BSA	Totals
Tanganyika/Tanzania (cont'd)	T20SM												1	1
	B.Cub												18	18
	All	17	79	35	16	12	29	22	31	47	34	10	33	365
Thailand	T15	30	8	6										44
	T20	6	10	97	79	50	21	30	12	24	8	8	2	347
	T20SL								4					4
	B.Cub												99	99
	S.Cub												109	109
	All	36	18	103	79	50	21	30	16	24	8	8	210	603
Trinidad	T15	9	4	13										26
	T20			5	7								2	14
	T20SL								2					2
	B.Cub												2	2
	S.Cub												4	4
	All	9	4	18	7				2				8	48
Tunisia	T15	3	4											7
	T20		2											2
	All	3	6											9
Uganda	T15	36	18	8										62
	T20	4	8	33	18	20	22	46	47	42	43	18	26	327
	T20C				3									3
	T20J						9							9
	B.Cub												16	16
	All	40	26	41	21	20	31	46	47	42	43	18	42	417
Unknown - Americas	T20 only					2								2
Unknown - Caribbean	T20			4						2	2			8
	B.Cub												4	4
	All			4						2	2		4	12
Unknown - Continental Africa	T15	2												2
	T20		1	8	2		6			12	10	3	3	45
	B.Cub												5	5
	All	2	1	8	2		6			12	10	3	8	52
Unknown - Far East	T20 only		1	7			4		2		71	3	22	110
Unknown - India/Pakistan	T20 only			1										1
Unknown - Middle East	T20 only				1		2				1		5	9
Unknown - Other Overseas	T15	3	1											4
	T20	1	4	7	1	9	3	3		3	1	38	36	106
	T20C				1									1
	T20S							1						1
	T20SH											2		2
	T20SR											1		1
	B.Cub												10	10
	S.Cub												3	3
	All	4	5	7	2	9	3	4		3	1	41	49	128
Uruguay	T15	2												2
	T20			1			10	4	2					17
	T20C						1							1
	T20SM												1	1
	B.Cub												1	1
	All	2		1			11	4	2				2	22
U.S.A. - Johnson Motors	T15	492												492
	T20	308	528	462	478	173	197	342	33	191	33	24		2769
	T20C				529	515	1							1045
	T20J					1	58	38	12					109
	T20CA					2	153							155
	T20S						420	673	181					1274
	T20T								144					144
	T20W							2						2
	T20SL								200					200
	T20SS									325	250	25		600
	T20SH									1		1		2
	T20SC										1			1
	T20SR												385	385
	TR20									59				59
	TS20										329			329
	T20SM											773	1429	2202
	T20M												766	766

Country	Model	1954	1955	1956	1957	1958	1959	1960	1961	1962	1963	1964	BSA	Totals
U.S.A. - Johnson Motors (cont'd)	'Export'												188	188
	All	800	528	462	1007	691	829	1055	570	576	612	823	2769	10722
	Engines								1		2			3
U.S.A. - Other	T15	24												24
	T20	9	6	13	9	22	9	4	2	1		1	1	77
	T20C				1	3	2							6
	T20S							1						1
	T20SL								1					1
	T20SH											2		2
	T20SR												3	3
	T20SM												6	6
	All	33	6	13	10	25	11	5	3	1		3	10	120
	Engines									2				2
U.S.A. - Triumph Corporation	T15	760	128	50										938
	T20	256	339	502	450	470	400	564	254	186	242	162	79	3904
	T20C				533	515	89							1137
	T20J				4	98								102
	T20CA				4	186								190
	T20S						428	567	123					1118
	T20T								150					150
	T20W							4						4
	T20SL								228					228
	T20SS									144	220			364
	T20SC									60	137	146	261	604
	T20SR									36	74	125	179	414
	T20SM											23	850	873
	T20M												464	464
	S.Cub												1	1
	'Export'												120	120
	All	1016	467	552	983	993	1201	1135	755	426	673	456	1954	10611
	Engines										1	1		2
U.S.A. - All	T15	1276	128	50										1454
	T20	573	873	977	937	665	606	910	289	378	275	187	80	6750
	T20C				1063	1033	92							2188
	T20J				5	156		38	12					211
	T20CA				6	339								345
	T20S						848	1241	304					2393
	T20T								294					294
	T20W							6						6
	T20SL								429					429
	T20SS									469	470	25		964
	T20SH									1		2	1	4
	T20SC									60	137	147	261	605
	T20SR									36	74	125	567	802
	TR20									59				59
	TS20										329			329
	T20SM											796	2285	3081
	T20M												1230	1230
	S.Cub												1	1
	'Export'												308	308
	All	1849	1001	1027	2000	1709	2041	2195	1328	1003	1285	1282	4733	21453
Venezuela	T15	18		4										22
	T20	2		17	28	19	24	38		6	12		6	152
	T20S							2						2
	T20SH												2	2
	B.Cub												6	6
	All	20		21	28	19	24	40		6	12		14	184
Virgin Is.	T15	1												1
	T20		1	3			2	8						14
	T20S							4						4
	All	1	1	3			2	12						19
West Irian	T15		2											2
	T20		10	16	11	4	1	1						43
	T20C					1								1
	All		12	16	11	5	1	1						46
Windward Is.	T20 only							2	4	4				10
Yugoslavia	T20 only		1		2									3
Zanzibar	T15		3	4										7
	T20		2											2
	All		5	4										9

Country	Model	1954	1955	1956	1957	1958	1959	1960	1961	1962	1963	1964	BSA	Totals
Total - All countries	T15	6240	2074	893	30									9237
	T20	1974	6446	7792	7714	7477	8923	10901	9572	4841	4696	3531	2721	76588
	ISDT				1									1
	T20C				1773	1872	466							4111
	T20J					5	166	38	14					223
	T20CA					14	351							365
	T20S						1074	2098	320				1	3493
	T20T						1	2	620	1			7	631
	T20W							6						6
	T20SL							1	1813					1814
	T20SS									576	553	56	8	1193
	T20SH									604	363	226	633	1826
	T20SC									60	137	153	261	611
	T20SR									36	74	133	571	814
	TR20									270	68	100	100	538
	TS20									24	340	1	2	367
	T20WD										4	88	2	94
	T20SM											801	3220	4021
	T20P												16	16
	B.Cub												1719	1719
	T20M												2106	2106
	S.Cub												2450	2450
	Tarbuk												46	46
	Export												402	402
	All	8214	8520	8686	9517	9368	10981	13046	12339	6412	6235	5089	14265	112672
	Engines	4	1					2	26	30	11	23	11	108

Appendix 7

Delivery details - by number of destination countries

Model	Destinations
T15	102
T20	142
ISDT	1
T20C	48
T20J	4
T20CA	8
T20S	42
T20T	15
T20W	1
T20SL	29
T20SS	37
T20SH	25
T20SC	2
T20SR	6
TR20	13
TS20	9
T20WD	6
T20SM	22
T20P	1
B.Cub	47
T20M	12
S.Cub	35
Tarbuk	1
Export	12
All	153

Published road tests

*(All USA fuel consumption figures have been converted to the equivalent UK Imperial gallon figure).

Date & machine details	Magazine	Speeds & acceleration	Fuel consumed	Both brakes
December 1953 1954 Terrier T15	*Motor Cyclist* (USA)	Top - 70mph 3rd - 56mph 2nd - 41mph 1st - 30mph	*108mpg average	N/A
June 1954 1954 Terrier T15	*Cycle* (USA)	Top - 68.44mph 3rd - 57mph 2nd - 41mph 1st - 26mph Standing $\frac{1}{10}$ mile in 11.17 secs. Standing $\frac{1}{4}$ mile in 19.88 secs.	*105mpg average	16ft 4in from 25mph
July 1954 1954 Tiger Cub T20 (T20 no.4955)	*Motor Cyclist* (USA)	Top - 76mph 3rd - 60mph 2nd - 44mph 1st - 32mph (Speeds obtained with 120 main jet and without air cleaner)	*103mpg average	
20th October 1955 1956 Tiger Cub T20 (T20 no.17513, SWD 193)	*Motor Cycling*	Top - 66.7mph = 43 secs. 3rd - 56mph = 18.2 secs. 2nd - 41mph = 7.6 secs. 0 - 30mph = 3 secs. 0 - 40mph = 7 secs. 0 - 50mph = 12 secs.	136mpg @ 30mph 112mpg @ 40mph 90mpg @ 50mph	29ft from 30mph
October 1956 1956 Tiger Cub T20	*Cycle* (USA)	Top - 66.66mph 3rd - 47mph Standing $\frac{1}{10}$ mile = 11.35 secs. Standing $\frac{1}{4}$ mile = 22.05 secs.	*Urban average 106mpg	80ft from 50mph
4th April 1957 1957 Tiger Cub T20 (T20 no.30244, VUE 547)	*Motor Cycle*	Top - 63mph 3rd - 55mph 2nd - 42mph 1st - 29mph Standing $\frac{1}{4}$ mile = 21.8 secs, reaching 56mph	130mpg @ 30mph 111mpg @ 40mph 90mpg @ 50mph	32ft from 30mph
October 1957 1957 Tiger Cub T20	*Cycle* (USA)	Top - 70.5mph 3rd - 52mph Standing $\frac{1}{10}$ mile = 11.0 secs. Standing $\frac{1}{4}$ mile = 21.5 secs.	*Test average 100mpg.	12ft from 20mph
23rd February 1961 1961 Tiger Cub T20 (T20 no.72269, 770 BUE)	*Motor Cycling*	Top - 64.3mph 0 - 30mph = 75 yards 0 - 40mph = 135 yards 0 - 50mph = c.500 yards	160mpg @ 30mph 96mpg @ 50mph 126mpg average	33ft from 30mph
Unknown date 1961 1961 Sports Cub T20SL (962 BUE)	*Motor Cycle Mechanics*	Top - 75mph 0 - 50mph = 10 secs.	81mpg average	32ft from 30mph
27th April 1961 1961 Sports Cub T20SL (962 BUE)	*Motor Cycling*	Top - 77.8mph 0 - 30mph = 50 yards 0 - 40mph = 100 yards 0 - 50mph = 200 yards	114mpg @ 30mph 84mpg @ 50mph 91.4mpg average	31.5ft from 30mph

Date & machine details	Magazine	Speeds & acceleration	Fuel consumed	Both brakes
14th February 1962 1962 Sports Cub T20SH (T20SH no.84011, 210 CNX) (Factory records show no.84011 as a T20)	*Motor Cycling*	Top - 79.2mph 0 - 30mph = 50 yards 0 - 40mph = 100 yards 0 - 50mph = 190 yards	112mpg @ 30mph 86mpg @ 50mph 92mpg average	31.5ft from 30mph
3rd January 1963 1963 Sports Cub T20SH (T20SH no.89010, 618 EUE)	*Motor Cycle*	Top - 74mph 3rd - 66mph 2nd - 42mph 1st - 28mph Standing ¼ mile = 20 secs. reaching 63.8mph	118mpg @ 30mph 100mpg @ 40mph 86mpg @ 50mph 72mpg @ 60mph	32ft from 30mph
14th April 1966 1966 Bantam Cub T20B (T20B no.129, GWD 434D)	*Motor Cycle*	Top - 65mph Standing ¼ mile = 22 secs. reaching 55mph	121mpg @ 30mph 118mpg @ 40mph 96mpg @ 50mph	33ft from 30mph
June 1966 1966 Bantam Cub T20B (T20B no.129, GWD 434D)	*Motor Cycle Mechanics*	Top - 68mph 3rd - 57mph 2nd - 36mph 1st - 16mph 0 - 60mph = 23 secs.	110mpg average	30.5ft from 30mph
9th March 1967 1967 Super Cub T20B (T20B no.4832, HOX 50E)	*Motor Cycle*	Top - 66mph 0 - 30mph = 5.5 secs. 0 - 40mph = 10 secs. 0 - 50mph = 16 secs. 0 - 60mph = 25 secs. Standing ¹/₁₀ mile = 21.2 secs. reaching 56mph	121mpg @ 30mph 115mpg @ 40mph 95mpg @ 50mph	31ft from 30mph

Appendix 9

Chronology of specification changes

DATE	MACHINE	MODEL	DETAIL
Oct? 52		T15	First prototype T15 Terrier constructed.
Nov 52		T15	Six pre-production prototypes built for testing and evaluation, three were Lucas equipped and three had Wipac electrics. The rear light was a Lucas MT110 type. Some of the earliest production machines had a Lucas L525 unit but this was soon replaced by the Lucas L529, without a stop-light. The rear number plate was shaped at the top for the L529 light unit and to form a lifting handle. Two small reflectors were used as the lower number plate bolts. The new model, an 'amended' prototype with Wipac electrics, was exhibited on the Triumph stand at the Earls Court Show. Delivery promised for June 1953.
May 53		T15	First gear indicator fitted to some prototype machines.
Jul 53	101	T15	New model. See Model Profile in App. 1. First machine in the factory despatch records - no delivery details. Single row roller big end fitted.
	102	T15	24/7/53 - first Terrier despatched from Meriden - to Australia.
Aug 53	101	T15	1954 Production year began.
	107	T15	19/8/53 - First home delivery to J.A. Hitchcock, Folkstone, registered RKO 88.
Sep 53	128	T15	The first few Terriers leave Meriden for the USA.
Oct 53		T20	New model, the 'Tiger Cub', announced in the UK press.
Nov 53		T20	The new model exhibited at the Earls Court Show.
Dec 53?	?	T15	An oil seal has now been fitted at the engine mainshaft where there was none before. The drive side inner cover altered to make provision for the seal.
Jan 54	1836	T15	Double filament rear light bulb introduced for use with a stop light switch.
	2278	T15	Frame - centre stand lug strengthened after breakages reported. Centre stand shortened to reduce stresses.
Mar 54	3000	T20	New model. See Model Profile in App. 1. No delivery details are shown for this first Tiger Cub machine. The T20 was fitted with a 1¹/₈in steel-backed white metal plain bush big end from the start. Tiger Cub mudguards had a raised central rib whereas the Terrier guards were plain in section.
	3001	All	Oil pressure test hole, closed by a screwed plug, added to front of crankcase.

DATE	MACHINE	MODEL	DETAIL
Mar 54	3001	All	The twinseat now has white piping around the top edge. It was previously all black and without any piping. An optional prop stand now seems to be available. Earlier frames seem to have been made either with or without the prop stand lug.
	3002	T20	The first Cub for which any delivery details are shown - to Zurich, Switzerland, 10th March 1954.
Apr 54	3905	T15	Big end bearing changed from a $1^1/2$in roller type to the Tiger Cub $1^1/8$in plain bush. New connecting rod for the new big end bearing. The crankpin has a $^3/4$in shoulder into the flywheels.
		All	New drive side outer cover with oil filler screw cap, sited above the alternator. Previously there had been no oil filler in the outer cover but a filler plug in the top of the drive side inner cover. This plug has been removed. Gear indicator finger now secured to the central spindle by a screw. Before that it had a push-in fitting. Clutch bearing is now a plain cast iron slipper ring - replacing the 58 ball bearings in a pressed-up housing. New clutch locknut. Oil pump feed plunger diameter increased from $^3/16$in to $^1/4$in for use with the new plain big end. The drive side engine mainshaft now secured in the flywheels by a serrated flange - replacing the earlier keyway and locknut. It is thought that the serrated type mainshaft was used on the T20 from the start.
	4127	All	Rocker box covers now made from a heavier gauge metal. Klingerit gasket material used.
	4859	All	Sludge trap now incorporated in timing side flywheel.
May 54	5360	All	A clutch adjuster and locknut now fitted in the pressure plate. Heavier duty clutch cable now used.
Jun 54	6670	All	Centre stand - new component with forged legs replaced the earlier tubular type.
Jul 54	7320	All	Wiring - battery becoming discharged due to short circuits in the stop-light cable - cables re-routed. PRS8 switch - the link between terminals 5 and 6 removed to increase the alternator charging rate at low speeds.
	7379	All	Crankpins - breakages reported due to incorrect material. New material now used - these crankpins identified by a number '6' stamped on the end. There was also an increase in the radius of the crankpin shoulder from 0.025in/0.015in to 0.055in/0.045in. The corresponding flywheels (E3336, E3360, E3419, E3424) had an increased chamfer at the crankpin hole from 0.040in/0.030in to 0.060in/0.050in.
	7578	All	Longer oil tank breather tube.
	8022	All	Clutch bearing - a porous bronze ring replaced the earlier cast iron type.
	8120	All	Leakages reported from the drive side and timing side outer covers after polishing. Special check on their flatness. Heavier engine outer covers to be introduced. Gasket material for outer engine covers improved. High gear oil seal - new $^5/32$in thick felt seal fitting more tightly on the gearbox mainshaft after leakages reported. Fibre washer material changed on the level and drain plugs after failures had been reported. Alternator rotors working loose and causing engine knocking at low speeds. Clearance decreased between rotor and crankshaft and a new lockwasher used. Wiring - cables being pulled out of the ignition/lighting switch. Harness modified with two clusters of snap connectors in the nacelle and a clip installed near the switch now prevents the wires pulling out. Wiring - snap connector fitted into the battery negative lead. Brakes - rear brake anchorage plate failures due to the peg in the left fork end coming loose. New screwed-in peg to replace the earlier pressed-in type. Kickstart failures due to the spindle housing in the timing side inner cover breaking away. Dimensions modified and inspection increased.
	8165		New clutch operating lever with increased leverage.
	8166		Crankcase oil leaking into the outer timing side cover through drain hole in the crankcase - hole repositioned.
	8259		Phosphor bronze clutch bearing ring fitted with 20 x $^1/8$in balls. Clutch sprocket hardened. These sprockets are stamped 'H' for identification.
	8364		Brakes modified (oval shaped) rear brake cam and spindle replaced the earlier more square-cornered type. New rear brake operating lever.
	8434		New gear change quadrant - quadrants reamed through and the plunger material changed.
Aug 54	8518	All	1955 production year began.
		T20	Exhaust pipe - a low level system replaced the earlier high level type which was retained as an option. Piston rings - chrome plated top compression ring used for the first time. New distributor for the T20 - Lucas 40433A. The T15 retained Lucas 40399B.
		T15	Exhaust pipe - continued to be low level but with a high level option for the UK. Drive side outer cover - oil filler screw finally sited at clutch centre. T15 now used the same cylinder head as the T20. It is thought that the T15 head did not have the enlarged holes to receive the cylinder barrel dowels as the T15 barrel was spigotted into the head.
		All	Prop stand lug now fitted to all frames. The prop stand itself was optional. This feature may have been extant from machine number 3001, or even earlier. The date of this improvement is uncertain. Rocker spindles - additional thrust washer. Two now fitted, one each side of the 'Thakeray' spring washer. Rectifier, Lucas 47111A, 'Sentercel' selenium type - $2^3/4$in diameter replaced the $4^1/2$in diameter germanium type. Improvements in the coating process of the semi-conductor material permitted a reduction of the carrier plate diameter. Alternator charging rate increased - see also no.7320. Frame - strengthened boxed crown lug - (steering headlug casting). Rear suspension plunger guide rod lengthened from $9^3/16$in to $9^1/2$in and a locknut fitted on the lower end. Amal 332 carburettor - choke removed. Gearbox - pawl retaining disc placed between the layshaft low gear and the kickstart spindle. Clutch - new shock absorber rubbers and spider.
	8730		Gearbox - layshaft blanking discs have been coming loose - new domed disc now used.
	10197		Oil pump - auxiliary balls fitted between the pump and the crankcase body as non-return valves to prevent the crankcase flooding when the machine was left for long periods.
Feb 55	11621	All	A single Lucas RER 16 rear reflector replaced the earlier twin D274 types. New clutch friction material, 'Neo-Langite' brought into use on machines nos.11621 - 11822, 12152 - 12222 and then

DATE	MACHINE	MODEL	DETAIL
Feb 55	11621	All	no.12704 onwards (March 1955).
			New clutch bearing ring with 16 ball bearings replaced the older 20-ball type on machines with the Neo-Langite plates.
			Machines with this type of clutch had a new drive side inner cover with a rubber washer clutch oil seal. Previously there had been no clutch oil seal.
			Clutch drive and driven plates - now three of each. The old type of clutch had two of each plate.
Mar 55	12704	All	Clutch bearing bronze ring had 16 x $^5/_{32}$in ball bearings.
Jun 55	15292	All	New clutch driving pins, made from improved material.
			New clutch driving cups machined from the solid, previously they had been pressings.
	16515		Clutch driving pins now riveted over the clutch backplate.
			New 'keyed' clutch locknut and tab washer. Previously only a spring washer had been used under the nut.
	17258		Oil pump feed hole elongated.
			Nacelle top - new type with separate cut-outs for the front brake and clutch cables which used to feed through the handlebar cut-outs.
			The options for saddle and pillion seat were dropped at the end of the 1955 season.
Aug 55	17389	All	1956 production year began.
			Frame modified at the rear to give extra tyre clearance.
			Oil tank capacity increased from $2^1/_4$ pints to $2^3/_4$ pints. Oil tank filler cap repositioned from the front corner of the tank to the top centre.
			New crankcase oil scavenge pick up pipe.
			New oil pump with increased output reported in the motor cycling press. No evidence of any change in the Parts List.
			Handlebars - new, slightly more upswept bend (still with welded-on lever brackets for the home market). Longer handlebar levers on T20 only. They were rather more pointed than the earlier type retained by the T15.
			Exhaust pipe - high level pipe option now for the USA only.
			New, heavier drive side and timing side outer covers. More chain clearance in the drive side outer cover. Drive side inner cover with threads ($^5/_{16}$in x 22TPI) tapped into two holes to facilitate its removal using puller D342T.
			Wiring harness - alternator mid-green wire given a yellow tracer.
			Wheels - butted 10/12 guage spokes used on the drive side of the rear wheels.
		T15	Petrol tank - mounting 'ears' extended right through the tank as internal strengthening braces.
			This tank was to be used for most sports Cubs from now on. The internal bracing now to be used on all tanks.
			Contemporary reports say that the T15 now had a deeper section chainguard but there was no change in the Parts List.
			'Long' type ($4^7/_8$in) cylinder barrel studs now used on the Terrier. Previously the 'short' type ($4^{19}/_{32}$in) used. The Terrier now has the Cub cylinder barrel but with a smaller bore.
		T20	Wheels - both now 16in with 3.25in tyres. Previously they were 19in with 3.00in tyres.
			Chainguard - new deeper section. The design is specific to this year and model only.
			Primary chain - now $^1/_2$in x $^3/_{16}$in single row replacing the former $^3/_8$in x $^7/_{32}$in for 1954/5. The chain was endless and made to tighter tolerances than the rear chain.
			Clutch sprocket and housing now with 36T x $^1/_2$in pitch. Previously 48T x $^7/_{32}$in pitch.
			Engine sprocket now with 18T x $^1/_2$in x $^3/_{16}$in replaced the earlier 19T $^3/_8$in x $^7/_{32}$in item.
			New 17T gearbox sprocket (was 18T) and 54T rear wheel sprocket (was 48T) - to allow for the reduction in wheel sizes to 16in and the changes to the engine and clutch sprockets.
			HT coil now located under the seat, leaving its former position on the rear mudguard behind the gearbox.
			A plastic valance under the twinseat protected the coil and rectifier from the elements - for the 1956 T20 only.
			New 3 gallon petrol tank. Broader and flatter than the earlier $2^5/_8$ gallon 'peanut'-shaped tank.
			Tank badges - same two-piece design as before but with different backing strips for the new tank shape.
			New prop stand and centre stand. Made shorter to make allowance for the 16in wheels on the T20.
			New speedometer 80421/36/2 (mph) and 80421/37/2 (kph) for the new rear wheel size.
			T20 nos.17389 to 18596 used flywheels that had a thickened outer rim but still retaining the smaller $^3/_4$in shouldered crankpin.
			Prior to that the flywheels had been the same thickness across their whole diameter.
			Rear mudguard - two curved brackets fixed to the mudguard for fitting it to the frame.
Oct 55	18597	T15 T20	Some flywheels made heavier by thickening the outer rim but still with the $^3/_4$in spigoted crankpin.
			New crankpin with $1^1/_8$in diameter at the big end and a $^{13}/_{16}$in spigot.
Jan 56		T20	Gearbox ratios - 'Works' trials machines were fitted with wide ratio internal gearing - a 29T high gear and a 16/25T layshaft cluster. (Later this became the normal WR gearset.)
Feb 56	22117	T20	New distributor - Lucas 40529A - 12° range.
	22464	All	Duplex primary chain kit (Part No. CD 112), available from June 1958 to convert any Cub after this number.
Apr 56	24090	All	Big end - Vandervell VP3 copper/lead alloy bush material replaced the previous white metal type.
			The timing side main bearing bush also in Vandervell VP3 replacing the earlier white metal material.
	24244		Drive side inner cover - increased chain clearance.
Jul 56	26106	All	Gear indicator - new type brought into use. It is not known whether this was the change to the printed dial from the earlier embossed type or whether the change was a new type of rack and pinion mechanism (or both)?
Aug 56	26276	All T20	1957 production year began.
			Frame, new swinging arm type with a single strengthening gusset under the headstock.
			The frame was in three parts: main loop, rear sub-frame and swinging arm.
			The rear suspension was controlled by separately bolted-on, non-adjustable, hydraulically damped spring units.
			Swinging arm bushes in Vandervell VP3. (Same material as the timing side main bearing and big end - see no.24090).
			Longer centre stand and prop stand - to suit the new frame.
			Rear mudguard - new item for the new frame.
			New silencer, slighter larger barrel than before with repositioned bracket for the new frame.
			Speedometer cable length increased for the swinging arm frame.
			Speedometer cable drive, previously over the wheel spindle, was now taken from under the spindle.
Aug 56	26276	T20	Exhaust pipe was the low level type only from now on. High level option deleted.

DATE	MACHINE	MODEL	DETAIL
Aug 56	26276	T20	Chainguard. Now in two parts - front and rear sections.
			Restyled twinseat, broader at the nose than before and with a new pan. A safety strap became optional.
			Cylinder head - still the 'Round' type, but the inlet rocker box now had finning for improved cooling.
			Front forks (lightweight) - now with hydraulic oil damping, replacing the earlier grease lubricated undamped fork.
			New rubber oil seal replaced the old felt seal. Chrome plated oil seal holders - previously cadmium plated.
			The stanchion diameter 0.010" less than the old type due to the better finish needed for the new fork seals.
			New front fork internals - springs, stud, top bush and restrictor rod.
			New middle lug and stem with smaller holes for the new stanchions.
			The top yoke remained unchanged because the tapers took up the difference in stanchion diameter.
			Wheel rims - centres no longer painted. (n.b. - This change may have taken place late in the 1956 year).
			Rear wheel now strengthened with butted spokes on both sides.
			Brakes - new rear brakeplate assembly. New malleable cast iron rear brake pedal replaced the earlier pressed type and a new brake pivot spindle. New rear brake rod for the new frame - approximately $17\frac{1}{2}$in long. The plunger frame rod was $17\frac{3}{4}$in long.
			Rear wheel adjusters - new type for the swinging arm frame with a smaller plate and less offset on the drawbar.
			Tapered piston ring, marked 'Top', now used as the middle compression ring.
			Primary chain tolerances tightened - reduced from 0.15% to 0.05% on nominal length.
			New drive side inner and outer covers.
			Oil tank - drain plug now fitted.
			Oversize ($1\frac{1}{4}$in) inlet valve now available in the USA. For use with cylinder heads before engine no.45086.
			Racing kit available in the USA?
	26500		Chrome top piston ring discontinued. Return to the normal type. Piston rings repositioned.
		USA	Rear suspension units - literature suggests that the American market models had three-position adjustable units.
			Tyre inflators no longer furnished as standard equipment in this market.
Jan 57	27660	T20C	New model. See model profile in App. 1. First T20C was sold to the USA.
			The new rear chainguard for the T20C had a main section similar in design to the old plunger frame type but with an additional front section (as T20). It became the standard item for all sports Cubs up to the end of 1964.
			The petrol tank was the T15 type which became the normal tank for all sports Cubs, except the T20SH.
			Upswept exhaust pipe, routed outside the rear sub-frame. This system was used on many later sports models.
			Rear sub-frame - new part with modified cross bracing to give additional clearance for the mudguard. The sub-frame right hand tube to the swinging arm was straight all the way and carried a bracket for the high level exhaust system.
			New oil tank with a froth tower. This type of oil tank was to be used on all future sports models.
			Rear suspension units were longer than on T20. These became the normal unit for many later sports Cubs and for the later 17in wheel T20 from no.56360.
			The rear suspension now had reinforcing brackets for the rear unit attachments to the swinging arm. These brackets were used on all subsequent sports Cubs.
			The front fork stanchions were 1in longer than those for the T20. These stanchions will also be used on the T20CA.
			A torque stay (from engine to frame top tube) was available in the USA.
			A new left side footrest bracket (marked with a 'C') had the prop stand at a steeper angle to allow for the larger wheels on the T20C. This left bracket will be used for all sports Cubs from now on.
			A new middle lug and stem and top yoke. Both items were standard T20 components but with added reinforcing.
			Wheels - 10/12 gauge butted spokes used front and rear.
			New front number plate with a small cutout for the top mudguard stay - later used on most sports Cubs.
Apr 57	31463	T15	The last Terrier - in a batch of 27 Terriers sent to Indonesia on 11th April 1957.
May 57		All	Riders Manuals published up to September 1955 referred to the camwheel as having three keyways (presumably for alternative valve timing options). By the 2nd edition of Manual no.4 (May 1957) this erroneous reference to alternative keyways had gone.
Jun 57	33340	All	Front forks - a $\frac{3}{32}$in hole drilled in the fork stanchions just above the bottom bush - to avoid build-up of pressure.
Jul 57	34214	All	New front fork seal holder. The 0.010in shim now no longer required. Having reduced the seal housing diameter by 0.010in, increasing the pressure on the oil seal and making it seal better, shim H1221 became superfluous.
Aug 57	35847	All	1958 production year began.
		T20	New slightly larger silencer with a revised footrest bracket gave a deeper note. (n.b. the old type silencer was retained on the T20C.)
			New more deeply valanced rear mudguard - unique to the 1958 T20 model.
			18T gearbox sprocket to suit the new lower geared primary drive. The sprocket shank diameter increased by 0.067in to fit inside the new oil seal.
		All	Duplex clutch sprocket, 48T x $\frac{3}{8}$in x $\frac{7}{32}$in with engine sprocket of 19T x $\frac{3}{8}$in x $\frac{7}{32}$in.
			Engine mainshaft lengthened on the drive side to make allowance for the duplex chain.
			The drive plate diameter for duplex clutches was 0.010in smaller than the simplex type.
			Primary chainline became $2\frac{21}{64}$in from conrod centre line. The simplex chainline figure had been $2\frac{3}{32}$in
			New distance piece between the alternator rotor and the engine sprocket.
			High gear seal - a single garter seal replaced the earlier felt type.
			Longer high gear bush now extended through the clutch oil seal.
			Deeper section inner and outer drive side covers to allow for the duplex chain.
			Clutch oil seal - a new rubber 'labyrinth' type with a cupped retainer replaced the earlier plain rubber seal and washer.
			Strengthened left footrest.
			New contact breaker cover clip. The new one did not have the pronounced 'shoulders' of the earlier type.
		USA	A Racing kit was available in the USA which included a T15GP carburettor and rubber mounted remote float chamber, light alloy valve spring caps, finned alloy rocker covers, racing clutch springs, tachometer, oil filter kit and a flywheel magneto.
			Tyre inflators now listed again as standard equipment.
	36094	T20C	New front frame with steering lock housing. The lock itself was optional.
			New middle lug and stem. Now with a slot to locate the steering lock tongue.
			The nacelle lower half now had no 'legs'. PVC gaiters now protected the fork seals.
	36350	All	The T20C front frame with steering lock housing now used on all models. There were a few later machines that did not

DATE	MACHINE	MODEL	DETAIL
Aug 57	36350	All	have the lock housing as the stock of old frames was used up. New middle lug and stem with a slot to locate the steering lock tongue. The lock itself was optional.
	36365	T20J	New model. T20 'Junior'. See model profile in App. 1. Identical to the T20 model but for the addition of a carburettor restrictor which brought the power output down below 5bhp, as required in some states of the USA.
	37220	All	The steering lock housing provided with a rubber cover.
Dec 57	38130	T20	New tank badges - the one-piece 'Mouth Organ' replaced the two-piece 'Four Bar' variety. New petrol tank of the same 'Flat' shape as before but with four threaded inserts to take the new badge instead of the earlier welded-on brackets. (n.b. Some machines after no.38130 exported to the USA still bore the earlier two-piece badge. The last known of this variant is no.38448.)
		All	New twistgrip - Amal 366/2 replaced the Amal 306 on all Cub models.
Jan 58	39167	All	New carburettor - Zenith 17MXZCS5, replaced the Amal 332. No insulating block between the carburettor and the cylinder head. Racing kit available in the USA including Amal 276 carburettor with rubber mounted remote float chamber, rear-set footrests and the $^{1}/_{2}$in pitch engine and clutch sprockets.
Feb 58	39733	T20CA	New model. See model profile in App. 1. Most machines of this type were sent to the USA .
May 58	42865	All	Crankcase - a cut-away in the casting now made provision for the fitting of a larger gearbox sprocket.
	45086	All	New cylinder head E3662 'Oval' type replaced the earlier 'Round' type. The casting will soon be altered to allow for a greater inlet port bore, if required.
Jun 58		All	Duplex primary chain kit (Part No. CD 112) available to convert any Cub after no.22464 from simplex to duplex chain.
Aug 58	45312	All	1959 production year began. New cylinder barrel, 'Oval' type with cast-in tunnels hiding the cylinder barrel studs. Replaced the 'Round' variety. A more substantial base flange was soon to come - (see no.45621). Valve guides shortened from $^{41}/_{64}$in to $^{9}/_{16}$in above the circlip - allows greater rocker movement with the new 'R' cam. Two thin oil seals now used at the high gear instead of one thicker type. Primary chaincase oil level reduced from 300cc to 200cc. An exhaust pipe option was a 10in long extension pipe in place of the silencer. For use in high performance work.
		T20	New 3 gallon petrol tank. The new tank was taller at the rear giving it a 'humped' appearance and retaining the one-piece badge. The badge was slightly different to the 1958 edition, allowing for a different side curvature of the new petrol tank. Semi-enclosure (side panels) now fitted, partly covering the rear of the machine. Oil tank and battery box moved rearwards and given longer bottom mounting bracket. Oil tank filler cap now sited at top/front of tank in its original 1954 position. (It had been centrally placed from no.17389.) The oil tank drain plug no longer fitted. New three-quarter length rear mudguard now finished under the seat, replacing the earlier full length variety. The rear mudguard was still ribbed in the centre. There was now a separate inner mudguard in sheet steel behind the gearbox. The air cleaner now no longer inside the battery box but separately bolted to the back of the carburettor. New silencer, 'fatter' than before and with a larger diameter exit pipe.
		T20C	The front fork felt washer above the oil seal now omitted.
		T20CA	Road tyres - Dunlop Universal - instead of the Trials Universal previously used.
	45621	All	The 'Oval' cylinder barrel now had a reinforced, thicker base flange.
Sep 58	45997	All	Gearbox mainshaft cluster amended to incorporate split rings.
	46130	All	Piston chamfer amended from 45° to 10° to conform with a cylinder head modification - see no.46179.
	46179	All	Cylinder head chamfer is amended to 10° to suit the piston - see no.46130.
Oct 58	46335	All	Cylinder barrel bore finish amended to 15 - 25 micro-inches.
	46621	T20	USA only - insulating block fitted between the carburettor and the cylinder head.
	46832	T20CA	An insulating block added between the Zenith carburettor adapter and the cylinder head.
Nov 58	47296	All	A counterbore added to the timing side inner cover to accommodate a tachometer drive.
	47301	All	A carburettor insulating block now fitted to all models for all markets.
	47348	T20J	Carburettor, a 9.5mm choke Zenith MXZ, was available for this model.
	47795	All	New alternator rotor nut.
	48139	All	No. 1 keyway in the drive side mainshaft modified to 84°. No information available on the earlier figure.
		T20S	New model. See model profile in App. 1 for details. Three basic specifications - Standard, Trials or Scrambles. In general the new model was much like the T20C except for certain performance parts and stronger front forks. This was the first model to use the Monobloc carburettor. Also used for the first time was the Energy Transfer ignition system. A Lucas RM 13 3-wire alternator and a 2ET coil. The alternator stator had two pairs of coils, one pair for ignition and the other for lighting. The ET distributor had a new cam with a very short (30°) opening period and 12° of advance. The rotor was the standard RM13 item but used the second (7o'clock) keyway on the drive side mainshaft. The new 'R' camshaft was fitted to many sports models from now on, had more dwell and 0.064in greater lift than the standard cam. Both the standard and the sports camshaft now had a tachometer drive slot machined in the outer end. The 'R' cam was used with interference type valve springs. Wide ratio and extra-close ratio gearbox clusters were also used for the first time on a production model. A new cylinder head with the larger inlet valve came into use. The frame main loop, F4544, had a bracket welded to the top tube 6$^{1}/_{4}$in forward of the centre line of the seat tube, on the drive side, for the remote float chamber to be fixed. The model was also the first to have heavyweight front forks, with a new front wheel spindle and brakeplate. The front forks were carried by a new heavyweight middle lug and stem and top yoke. It is believed that the rear sub-frame for the T20S and some of the later models had a right hand tube (carrying the high level exhaust bracket), with an outward bend at the bottom. The part number of this sub-frame (F4253) was the same as the earlier type with the straight right hand tube.
	48340	All	New drive side engine mainshaft incorporating two keyways. The no.1 keyway (at 3 o'clock) was for the T20, T20C & T20S high compression models with 16° BTDC static timing. The no.2 keyway (at 7 o'clock) was for T20S low compression models with 8° static timing. (n.b. the Triumph Corporation (USA) service bulletin says to use either keyway for battery ignition models.)

DATE	MACHINE	MODEL	DETAIL
Jan 59	48937	All	Machining deleted from drive side inner & outer cover screw holes. Die modified for the inner cover.
	USA	T20	The seat for the JoMo model given a safety strap.
Feb 59	USA	T20S	USA scrambler version - a 200 size carburettor main jet was an alternative for the normal 140 size. The fork covers had no headlamp mounting brackets. No headlamp, rear light, horn, horn push or number plates. No wiring in the harness for head or tail lights. Front mudguard no longer drilled for number plates. Rear reflector fitted to JoMo machines only.
	USA	T20CB	New model? This model was described in various factory memoranda as a replacement for the T20C. Its specification was exactly as for the T20CA except for an upswept exhaust system. Its colours were to have been Aztec Red and Black. None appear in the despatch records and it is assumed that none were built or sold - probably due to the recent arrival of the T20S.
	50051	All	Piston - valve cut-away increased from $^5/_8$in to $^{11}/_{16}$in. Gearbox mainshaft and layshaft assemblies redesigned for greater strength. Made from stampings. Oil pump skew gear - new material - cast iron - replaced phosphor-bronze. Oil tank filter - gauze amended from 60 mesh to 40 mesh.
	50459	T20	Petrol tank internal anti-surge baffles added.
Mar 59	50825	T20	A mute added to the silencer, secured with a self-tapping screw. The mute became optional on the other models.
May 59	52974	T20C	The last T20C, despatched to Norway.
Jun 59	53780	T20CA	The last T20CA, sent to the USA.
	56145	T20T	New model. See model profile in App. 1. Prototype, first machine, sold second-hand on 31/10/61. The T20T was in production for 1961 only - see no.64591 (apart from a few machines in some other years).
Aug 59	56360	All	1960 production year began. New swinging arm with the ability to take a 4in tyre. Wider cast lug and a straight right side arm. Previously both arms had been curved. All models now had the longer rear suspension units formerly fitted to the sports models only. Cylinder barrel studs shortened from $4^7/_8$in to $4^{19}/_{32}$in. The cylinder head could now be removed without lowering the front of the engine. Cylinder head nut length increased from $^9/_{16}$in to $^{11}/_{16}$in for use with the shorter barrel studs. New 'Oval' cylinder head with larger inlet port and inlet valve. New cylinder head was E4049A with a cast-on number E4102. This had a greater counterbore depth (approx. $^7/_{16}$in) for the cylinder head nuts than the earlier 1959 head E4049 which had a cast-on number E3662 and a counterbore depth of approx $^1/_4$in. The inlet valve diameter increased from $1^1/_8$in to $1^5/_{16}$in on all models. The cylinder head exhaust port moved around by 7° to give more frame clearance for the exhaust pipe. New pushrod tube - still had the guide plate inside but its locating slot and the peg in the crankcase were now omitted. The tube was now positioned by means of markings on the tube and rocker box. 'Langite' now used as the clutch friction material - replacing Neo-Langite. Or was this just a new name? New contact breaker drive shaft bush.
		T20	Zenith 18MXZ-C17 carburettor size increased to 18mm. A tickler now fitted plus adjustment for the pilot jet. The choke now became independent of the throttle action and had a black plastic operating knob. 17in wheels front and rear from now on - previously 16in. Dunlop tyres, 3.25in size, ribbed front and Universal rear. Taller centre stand made allowance for the larger wheels. New speedometer - SN.3153/17 (mph) or SN.3153/18 (kph). New rectifier, Lucas 47132B (square shape) replaced the old Lucas 47111A type, (circular $2^3/_4$in diameter). New contact breaker, Lucas 40699, replaced the old 40529A type.
		T20S	Tachometer option - driven from the end of the camshaft. Used a timing side outer cover incorporating the tachometer drive. This was the first production model to include a tachometer option. Monobloc 376/272 carburettor. (Zenith 18MXZ-C18 also listed.) The 376/272 now became the standard instrument for many sports Cubs. New rear mudguard - became the standard sports Cub blade. New front mudguard stays. New contact breaker - Lucas 40700. Replaced the old 40664A type. USA T20S machines did not have a tyre inflator for this year. The small triangular toolbox came into use.
Oct 59	57617	All	New two-piece crankcase with new oil scavenge pipe and oil filter. Plain bush main bearing still used on the timing side. New drive side outer cover. New timing side outer cover, now with a screw hole for an internal distributor clamp. Contact breaker internal clamp replaced the earlier external clamping plate. New frame front loop. Modified engine front mount allowed the engine to be fitted without the use of spacers. Press announcement - 'All sales of Triumph machines in France will be handled by Terrot. All machines supplied to France are to carry Terrot badges in addition to the Triumph motif'.
Jan 60	60265	T20W	New model T20 'Woods'. See T20S model profile in App. 1. This was an Enduro/Reliability Trial variant of the T20S trials version and was based on the factory's successful Scottish Six Day Trial machines. The records show only six machines sold but there were almost certainly more than that.
	60290		The last T20W shown in the factory despatch records. Delivered to the Triumph Corporation USA.
Feb 60	63233	T20	Oil filter - three experimental engines built with a paper element oil filter in the oil feed line to the crankshaft.
Apr 60	64591	T20T	New 1961 model pre-production machine. The first production machines came in August 1960. See also no.56145.
	64592	T20SL	New model. See model profile in App. 1. 1961 pre-production machine with a 1960 number. The first production machines were delivered in September 1960.
Jul 60	69157	All	Larger, brass-bodied oil pump. Modified porting to eliminate irregular wear on the ball valve seats. The plunger sizes were now $^5/_{16}$in, scavenge $^3/_8$in, the stroke was still $^3/_{16}$in. The new pump gave roughly 60% more throughput of oil. There was a new timing side main bearing bush to go with the new oil pump. It had a slot in the bush carrier sleeve to provide an oil bleed to the distributor skew gear. This type of bush was NOT be used with the earlier, small capacity oil pump as there would be insufficient oil to adequately feed both the skew gear and the big end.

DATE	MACHINE	MODEL	DETAIL
Aug 60	69517	All	1961 production year began.
		T20	A chrome styling strip now covered the seam between the top and bottom halves of the petrol tank.
			The tank badges had a small cutaway to make room for the chrome styling strip.
		Sports	The small triangular toolbox used on all the sports models, except the later T20SH which retained the battery box.
			All current sports models had Energy Transfer ignition systems with 12° of advance. (A contact breaker with 5° of auto advance was available in the USA.) Additional alternator coils for a stop-light, if fitted, on the T20T and T20SL. All sports models now had a tachometer option.
		All	Sales literature suggested that cork has replaced 'Langite' as the clutch friction material.
			Contact breaker cover secured by a self-tapping screw fitting into a spring steel bracket, replaced the earlier wire clip. New type of cover with a hole for the screw instead of a pressed slot for the wire clip.
			New crankcase oil scavenge pipe, end-cut at 90°, to replace the earlier 45° pipe.
	69708	T20T	The first production T20T machine built - delivered to Belgium in early September.
	69833	T20SL	The first production T20SL machine built - delivered to the USA in late September.
Oct 60	70805	T20J	The last T20J, delivered to Bermuda.
Mar 61	75973	T20S	The last T20S, delivered to Denmark. This model was dropped from the catalogue having been replaced by two new models for 1961, the T20T and the T20SL, the latter coming in Road and Scrambler versions.
Jul 61	78600	T20	New speedometer SN 3170/00 (mph) or SN 3170/01 (kph).
	80921	T20T	Forty machines delivered to the War Office which were painted Deep Bronze Green. This model was a modified T20T with a $^5/_{16}$in wide rear chain, gearbox and rear wheel sprockets instead of the usual $^3/_{16}$in size. It also had reinforced footrests, a low level exhaust pipe and the rocker oil feed pipe was $^3/_{16}$in, rather than the usual $^1/_8$in, diameter. This was the first sports Cub to have a centre stand as original equipment. A new rear chainguard with enlarged mudshield and additional protection for the lower chain. The model became the pattern for the later T20WD.
	81889	T20SL	Last T20SL - despatched to Eire in September 1961.
Aug 61	81890	All	1962 production year began.
			Lucas rear light unit L564 replaced the earlier L529. The rear number plate shape changed to suit the new light unit. No separate reflector from now on as the L564 lens had one built in.
			Cast iron oil pump. Same dimensions as the previous brass bodied type but gave better performance at higher temperatures. This pump also used stronger springs than before. A liner in the crankcase scavenge oil pipe (E4401) helped to prevent 'Wet Sumping' when the engine was run at high speeds in high temperatures and for prolonged periods.
			Pushrod cover guide plate now omitted. (The new tube has the same part number as the old one.)
			Some dealers had problems locating the pushrods during assembly. A simple forked tool (D618) was produced.
			Ewarts 504 'push-pull' type of petrol tap replaced the old 'taper-cock' type except for military models, including the much later French Army machines. This change actually took place only when stocks of the existing taps were exhausted. The machine number of the change is not known but it was some time between January and August 1961.
			New gearchange quadrant incorporating a keeper plate and screw.
		T20	Amal 32 carburettor being fitted to export T20 machines, with a dry felt air cleaner, (see also no.83192).
			Horn - Lucas 8H now standard on all the battery models, replacing the Clearhooter HF150.
			The Energy Transfer models have been using the Lucas HF1950 type since machine 48139.
			New type of chrome petrol tank styling strip for the tank seam.
			Seat top cover colour now usually grey but there may have been black topped seats as well. The T20 still used a single level seat.
		Sports	A two-level, grey topped seat is now used on these models.
			Gearbox - a 28T layshaft low gear became available for use with a 17/25 tooth mainshaft cluster converting the extra-close ratios to ultra-close, even more suited for racing purposes.
	82276	T20SS	New model. See model profile in App. 1. The first machine was sent to Eire in September 1961. The high compression version replaced the T20SL as the all purpose sports machine suitable for on/off road use. The low compression version replaced the T20T.
Dec 61		Sports	A new 10° range ET contact breaker (Lucas 40846) replaced the old 12° item.
Feb 62	82756	T20SH	New model. See model profile in App. 1. This was a pure sports road machine with a battery, lights and conventional ignition system. (This particular machine (no.82756) was eventually sold on 24/7/62.)
	83192	T20	The Amal 32 carburettor became available on home market machines.
	83727	T20	Lucas RM18 alternator replaced the earlier RM13. (The RM18 gave 42W @ 2000rpm and 57W @ 5000rpm.) It had a slimmer, larger diameter rotor and a new rotor sleeve, nut, lockwasher and distance piece.
	84269	All	New, two-piece crankcase, split as before at the barrel centreline but with a new ballrace timing side main bearing. This meant that the old drive side flywheel, now with a new timing side mainshaft pressed in, became the new timing side flywheel. There was also a new timing side inner cover. The oil pressure test hole was now moved lower down the crankcase.
			Separate oil pump skew gear and timing pinion replaced the earlier one-piece item. The skew gear was now made in steel, replacing the earlier cast iron type.
			The big end was now $1^5/_{16}$in diameter with a two-piece crankpin and new big end bush and conrod.
			New clutch shock absorber and larger threaded pins with larger threaded holes in the backplate.
		TR20	New model announced in the press.
		TS20	New model announced in the press.
Mar 62	84912	T20SC	New model. See model profile in App. 1. This was an export only (mainly the USA east coast), high performance on/off-road model. It was to all intents and purposes the 'scrambler' version of the T20SL, but with lights.
	84967	T20SR	New model. See model profile in App. 1. This was the 'on-road' version of the T20SC and, like the SC, it was also a USA east coast model (but several hundred were sold on the west coast too in 1965). The specification was much like the home market T20SH, but with ET ignition and extra-close ratio gears.
	85108	TR20	New model. See model profile in App. 1. This was a 'works replica' pure trials machine developed from the T20SS. It was the first production model to have the high level exhaust pipe inside the rear sub-frame. The rear sub-frame right hand tube was cranked outwards at the top and bottom. The footrests were a rearward facing and folding type later used on the Mountain Cubs.
			Brakes - the rear brake pedal pivot was moved backwards and the rear brake rod was only $15^1/_4$in long. The frame main

DATE	MACHINE	MODEL	DETAIL
Mar 62	85108	TR20	loop was double-gusseted under the headstock for additional strength. The new rear chainguard with enlarged mudshield and additional protection for the lower chain run had been borrowed from the military T20T model. Alloy mudguards were used which became alternative items for other models.
		T20SST	This model type sometimes appeared as part of the engine number on the first few TR20 machines, but they were shown in the despatch records as TR20.
Apr 62	85323	TS20	New model. See model profile in App. 1. This was the scrambles version of the TR20 and they shared many components.
	85698	T20T	The penultimate T20T. This machine had been sent on loan to Terrot, France and was probably the 'original' for the French Army Cub. It was later sold by Meriden to a Kent dealer in October 1964.
May 62	86204	Sports	RM19 Energy Transfer 4-wire alternator on all sports models except T20SH. The RM19 gave 50W @ 2000rpm and 60W @ 5000rpm. There was also a new Energy Transfer ignition coil - the 3ET.
Aug 62	88347	All	1963 production year began. Side points. The contact breaker with 10° of auto-advance was now driven from the end of the camshaft, replacing the old distributor. New crankcases - the timing side half reshaped internally to provide better draining to the scavenge oil pipe. New camshafts (standard and 'R' type) had a taper-bored outer end to carry the new type of auto-advance unit. The camshafts had an advance unit locating peg inside its tapered end so that the unit could be accurately positioned. New timing side outer cover with a 1³/₁₆in diameter opening behind the kickstart lever. Rubber grommet gave access to the clutch cable end making replacement of the cable much easier than before. New timing side inner cover and drive side outer cover. Finned alloy rocker covers replaced the old chrome plated pressed steel type. New Lucas 49072 silicon rectifier, 1¹/₂in diameter, black painted. Adjustable rear suspension units were optional on all models for this year only. Oil junction block - pipe material changed from copper to steel. New, grey faced speedometers SSM 2001/00 (mph), and SSM 2001/03 (kph). Used on all future models. New rear wheel speedometer gearbox and cable. A two-level seat with a grey top now used on all models.
		T20	New gear indicator, located in the top of the gearbox, replaced the indicator in the nacelle. New nacelle top cover carrying two separate Lucas 88SA switches which replaced the single PRS8. New headlamp light unit with a different bulb adapter and a new metal headlamp bulb holder. New wiring loom incorporating moulded connectors to the switches. New oil tank for the side panels T20 - now with a drain plug.
		Some	Exhaust pipe - low level. New exhaust pipe with a welded-on threaded boss for a stay to the front engine mount. Exhaust pipe - high level. New exhaust pipe going inside the rear sub-frame. New silencer for high level systems, angle-cut at the exit.
		Sports	The tachometer drive was now taken from the old distributor drive point behind the cylinder barrel plinth as the previous drive position at the end of the camshaft was now occupied by the contact breaker auto advance unit. New crankcase undershield for the T20SS and T20SH. Exhaust pipe - the high level exhaust pipe option was dropped for the T20SH.
Sep 62	88658	T20WD	New model - first machine. This machine was kept on Triumph company service until the end of 1964 when it was sold to a Lancashire dealer.
Jan 63	91085	T20	Radio pack fitted to Police machines for the Royal Ulster Constabulary. The three gallon petrol tank had four threaded holes in its top surface to secure a grid carrying the handset. The seat had a shallow section behind the rider for the radio power pack. There was a loudspeaker mounted on the handlebars.
Mar 63	92241	T20	New handlebar no longer had welded-on lever lugs but now with a bolt-on clutch lever and combined twistgrip and brake lever*, both of which were of the folded, rather than solid design. (*Home and general export machines only. The USA model retained the separate Amal 366 twistgrip and brake lever.) New brake and clutch cables.
Aug 63	94600	All	1964 production year began. The crankpin was still 1⁵/₁₆in diameter but now of one-piece construction. Oil pump skew gear drive material now 'Hidurax' aluminium bronze. New clutch pressure plate, cups and springs. The three clutch spring nuts should now protrude about ³/₈in proud of the pressure plate. Previously they would have been flush with the plate surface. Reshaped rubbers in the clutch shock absorber.
		T20	The rear mudguards no longer had a raised central rib, although the painted centre line and pin-striping remained. Optional front mudguards without holes for the number plate brackets as the legal requirement for this item had gone. Adjustable rear suspension units were shown in the USA east coast catalogue. Tyres - for the USA a ribbed front tyre was shown in the sales catalogue and a Dunlop K70 Gold Seal at the rear. New combined switch for the horn and dip (Lucas model 25SA) replaced the old horn push which screwed into the left handlebar (H886) and the Lucas 31549 dipswitch.
		Sports	The new rear chainguard with enlarged mudshield and additional protection for the lower chain run now in general use. Crankcase undershield - the sports models reverted to the 1962 type, except for the T20SH which lost this item. Adjustable rear suspension units were now standard on the T20SS, T20SC, TR20 and TS20.
		T20SS	Gone was the previous upswept handlebar option. 'Flat' handlebars only from now on for the T20SS and T20SH. The silencer was now the T20 type for the low level system or the TR20 type for the high level system. New front frame loop, double-gusseted under the headstock. Rear sub-frame right hand tube now cranked outwards at the top and the bottom making more room for the exhaust pipe on the optional 'inside' high level system.
	94600?	All	Main frame loop. There seemed to have been a change of policy with regard to the numbering of the frames from about machine number 94600. Previously the frame number had begun with 'T', followed by the machine number, but now the model type also appeared as part of the frame number; e.g. T20 xxxxx instead of T xxxxx.
Nov 63	95152	T20SM	New model. See model profile in App. 1. The Mountain Cub, intended primarily for the USA west coast. Model type redesignated as the T20M in November 1966. The carburettor on this model was a Monobloc 376/314.
Jul 64	99639	Fr. Army	See model profile in App. 1. The first French Army machine built. Delivered on 1st September to CGCIM, Paris. The model had several features derived from the earlier military T20T, (see no.80921). It also had footrests secured by a long

DATE	MACHINE	MODEL	DETAIL
Jul 64	99639	Fr. Army	bolt and nut rather than the more usual stud and two nuts. The rocker oil feed pipe was $^3/_{16}$ diameter rather than the usual $^1/_8$in.
	99719	T20WD	The last T20WD - sent to Paris for the French Army in September 1964.
		All	The decision is taken to move the Cub production line to the BSA works at Small Heath. Press anouncement not made until November 1964.
Aug 64	99720	All	1965 production year began - but see also 99733.
		Sports	Heavyweight front forks now with springs outside the stanchions whereas previously they had been inside.
			A splined kickstart lever, highly cranked and pivoting from near the shaft, replaced the cottered type.
			The rear sub-frame on all sports models was now cranked at the top and bottom on the right hand side.
			The front frame loop with double strengthening gussets at the steering head became standard on all sports Cubs.
			New swinging arm with an additional mounting point at the front for the lower run of the chainguard.
			New rear mudguard. The difference between this and the previous year's type is not known.
		Some	Lucas L679 rear light unit replaced the L564 type on the USA T20SS and TriCor T20 model.
		T20	No gear indicator at all from now on.
			Adjustable rear suspension units (Girling 640/545/20) option. Export models only?)
			Gearbox - close, extra-close and wide ratio gears available for the T20.
		All	The rear number plate was lengthened to accommodate a suffixed registration number. (This change may have taken place earlier as the first 'A' suffixed registrations came into use in January 1963.)
			'Deva-Metal' graphite impregnated sintered big end bush replaced the copper/lead Vandervell VP3.
			New clutch friction material - granulated cork with a synthetic rubber content.
			Sludge trap capacity increased. New, larger diameter screwed plug and a new timing side flywheel. New flywheel assembly incorporating the new timing side flywheel.
		T20SR	There were by now two versions of this model. The USA east coast model had the 9:1 piston and Energy Transfer ignition. The west coast model the 7:1 piston, AC/DC electrics, battery and battery box. Dunlop K70 Gold Seal tyres.
	99733	All	There is conflicting information as to the serial number of the first 1965 machine. The factory 'Build' book shows machine 99720 as '1965 model', and the illustrated Parts List no.10 begins with the same number signifying the start of 1965, but the factory 'Despatch' book has the comment '1st 1965 model' against machine 99733.
	101	All	The 'Oval' cylinder barrel and large valve cylinder head were still used on all models.
		Sports	The camshaft no longer had the advance unit locating peg.
			New 5-wire alternator stator (Lucas 47188) on Energy Transfer models.
			Contact breaker - side points type but now with 5° of auto advance (previously 10°) to cure high speed misfiring.
Nov 64	775	T20P	New Police model. As T20 but had a petrol tank with four screw holes for a radio pack. Also a loudspeaker mounted on the handlebars and a seat with a shallow section behind the rider for the radio power pack. The side panels were modified to carry extra switches and the exhaust system had a larger silencer with a built-in mute. The centre stand was reinforced by a gusset and there was a stop bracket and rubber stop grommet for the stand, a larger alternator, a larger battery and an altered battery box.
	833	T20T	The last T20T - delivered to Pakistan. The huge majority of T20T models were delivered in 1961, but one was built in 1962 and seven in the 1965 season, of which this was the last. It is not known whether these latter machines had distributor or side points engines.
		All	Press announcement - Cub production is to be moved to the BSA works at Small Heath, Birmingham.
Jan 65	1715	T20SS	The last T20SS. Delivered to Morison Jones, London, for export to the Far East.
Feb 65		All	The manufacture of parts for Cubs was now being carried out at BSA Small Heath works. The first complete machines were built at BSA in week commencing 13th August 1965.
	Various	All	New 'Square' cylinder heads and cylinder barrels. The barrel has eight fins instead of nine on the oval type. Short cylinder barrel studs for the 'Square' barrelled machines. New cylinder head nuts on 'Square' headed models - 1$^1/_{16}$in high.
	1867	TR20	The last TR20. Delivered to a dealer in Hampshire.
	1870	TS20	The last TS20. Delivered to Canada.
	2001	T20	The tyres were Dunlop Lightweight, ribbed front and studded rear.
Jul 65	2029	T20SR	The last T20SR. Sent to the Triumph Corporation, USA.
	2034	T20SC	The last T20SC. Sent to the Triumph Corporation, USA.
Aug 65	Various	All	1966 production year began.
		Some	Cubs being built at the BSA works Small Heath from 13/8/65. Frames were still temporarily being made at Meriden.
			The oil pump now driven by a slider-block mechanism similar to that employed on the twin cylinder models. The pump dimensions were the same as before but the new drive had an increased throw giving a further 50% delivery of oil. The new oil pump drive mechanism used a skew drive with a longer peg on the slider block and it had a narrower slot in the top for the tachometer drive. This new oil pump drive was not used on all Bantam Cubs - some early machines retained the old type of drive and skew gear. Oil pump auxiliary balls no longer specified.
		Sports	The cottered kickstart shaft and lever brought back into use on all sports models.
Sep 65	2619	T20P	The last T20P - delivered to the Chief Constable of Warwickshire.
Oct 65	2001	T20M	This model designation is now being used for the 'Mountain Cub' although T20SM will continue for several years more. See also no.3001/ Oct 66.
	4011?	T20	This is thought to be the highest numbered and the last Triumph framed T20 Cub to be built at the BSA works.
Dec 65	101	All	A single oil seal was now used at the high gear.
		B/Cub	New model. See model profile in App. 1. The Bantam Cub was a D7 Bantam with a Cub engine. New exhaust pipe for the Bantam Cub (also used for the later Super Cub). The lower run of this pipe had an outward bend, whereas the T20 pipe was straight at this point. A new silencer now had an entry pipe outboard of the silencer centre line whereas it was inboard on the T20. The tank badges on the Bantam Cub were the 'Mouth Organ' type but they differed from the T20 type in that they did not have a cut-away at the front or rear as the Bantam Cub tank did not have a seam along its side. The capacity of the oil tank now increased to 4 pints and the petrol tank decreased to 2$^1/_4$ gallons.
Feb 66	4373	T20SH	The last T20SH. Despatched to a Yorkshire dealer. There were later deliveries but this was the highest serial number.

DATE	MACHINE	MODEL	DETAIL
Aug 66	Various	All	1967 production year began.
			The big end changed from a plain bush to a caged double row needle roller. New conrod and crankpin to go with the new bearing.
			New camshaft assembly for all models (E7584 and E7585 with new pinion E7302) with the same profiles as before. The difference to the earlier type is not known for certain but may have lain in surface hardening.
			New oil pump assembly announced. This had the two plunger ends which located into the slider block, fixed together by means of a plate, instead of being separate.
			Gearbox - wide ratio and close ratio gears were now optional on the Bantam Cub. Extra-close ratios optional on all models.
			New piston rings for all models. The top plain and middle taper rings were now replaced by two new taper rings with a greater radial depth than the old 7:1 type. The new ring was used for both compression ratios and in both positions.
Oct 66	3001	T20M	The Mountain Cub continued. See T20SM model profile in App. 1. From October 1965 the 'T20M' designation was being used in sales literature and was stamped as part of either the frame number or the engine number. Machine no.3001 is believed to be the first with a 'T20M' designation stamped on both components.
	3333?	S/Cub	New model. See model profile in App. 1. The Super Cub was a D10 Bantam with a Cub engine. The model was also known in its early days as the 'T20 De Luxe'. It had the same exhaust pipe and silencer as the T20B Bantam Cub. The petrol tank capacity for this model was 2 gallons.
		T20B	Wipac electrics were now being used on many Bantam and Super Cubs as well as Lucas equipment.
Nov 66	3855	All	New gearbox mainshaft cluster and mainshaft low gear. Now with low profile 'stub' teeth which must not be mixed with the old style components.
Mar 67	7849	B/Cub	The last Bantam Cubs being built. This machine was delivered to Jamaica in December 1968.
Apr 67	8179	T20M	This was the highest numbered T20M shown in the records (apart from the French Army machines). It was delivered to Kenya in September 1967.
	9167	T20SM	The highest numbered T20SM in the records. Delivered in November 1967 to a Buckinghamshire dealer.
Jul 67	10050	Fr. Army	The last French Army Cubs being built and stored in warehouse. Deliveries continue spasmodically up to May 1970.
Aug 67		S/Cub	The Super Cub now fitted with a Concentric carburettor type 622/1. Cylinder barrels may have been painted black for the final year of production.
Jan 68	2422	T20B	'Tarbuk' conversion, first machine. New model? Sixty six machines from a cancelled export order to Iran were sold to Elite Motors, Tooting, London. See Chapter 6.8 for the known model details.
Mar 68	5668	T20B	The last of the 'Tarbuk' conversion machines.
Jan 69		T20B/175	New model? The first 8 machines of a type described as the 'T20 Bantam 175' were built. The model may have been a Bushman with Cub forks, tank and mudguards. See Chapter 6.9 for the known model details.
May 69		T20B/175	A further 240 machines of this type were built.
Jun 69	9710	S/Cub	The last Super Cub (from a batch of 53 machines). These were built in week ending 27/6/69 and were mostly delivered to France (painted in NATO Green) or to other export markets.
May 70	10050	Fr. Army	The last French Army machine in the despatch records. These machines had been built in July 1967.
Jun 70	7873	T20B	The last Cub of any kind to be delivered. It was a T20B machine, probably a Super Cub but the model type is not actually shown in the despatch records, and it was sent to Slocombes Ltd., London NW10, on 18th June 1970.

Appendix 10

Useful addresses

Tiger Cub and Terrier Register
Mike Estall
24 Main Road
Edingale
Tamworth
Staffordshire
B79 9HY
England
Phone: 01827 383415
Fax: 01827 383183

Triumph Owners Motor Cycle Club
Mrs Margaret Mellish (General Secretary)
4 Douglas Avenue
Harold Wood
Romford
Essex
RM3 0UT
England
Phone: 01708 342684

Triumph Owners International Club (USA)
PO Box 6676
Holliston
Massachusetts
USA
Phone: 800 451 5113
FAX: 508 429 4221

Triumph Owners Club Nederland
Erik Boelen
Strobloemstraat 57
5643 JW Eindhoven
The Netherlands
Phone: 040 112702

The Vintage Motor Cycle Club
Allen House
Wetmore Road
Burton-on-Trent
Staffordshire
DE14 1TR
England
Phone: 01283 540557
FAX: 01283 510547
Internet: www.motorcycle-uk.com

Appendix 11

Further reading

Triumph (Single Cylinder) - A. St. J. Masters. 1961. Published C. Arthur Pearson Ltd. (5th Edition).
Triumph Tiger Cub & Terrier - Pete Shoemark. 1978. Haynes Publishing Group. ISBN 0 85696 414 X
It's a Triumph - Ivor Davies. 1980. Haynes Publishing Group (Foulis). ISBN 0 85429 182 2
Classic British Trials Bikes - Don Morley. 1984. Osprey Publishing Ltd. ISBN 0 85045 545 6
Whatever Happened To The British Motorcycle Industry? - Bert Hopwood. 1984. Haynes Publishing Group (Foulis). (Hardback) ISBN 0 85429 241 1
(Paperback) ISBN 0 85429 459 7
Pictorial History of Triumph Motor Cycles - Ivor Davies. 1985. Temple Press. ISBN 0 600 35169 6
Illustrated Triumph Motorcycle Buyers Guide - Roy Bacon. 1989. Niton Publishing. ISBN 0 9514204 0 2
Triumph Singles - Roy Bacon. 1991. Niton Publishing. ISBN 1 85579 009 2
Classic Motorcycles, Triumph - Don Morley. 1991. Osprey Publishing. ISBN 1 85532 124 6
Triumph in America - Lindsay Brooke and David Gaylin. 1993. Motorbooks International. ISBN 0 87938 746 7
Practical British Lightweight Four-Stroke Motorcycles - Steve Wilson. 1994. Haynes Publishing Group (Foulis). ISBN 0 85429 901 7
Tales of Triumph Motorcycles & the Meriden Factory - Hughie Hancox. 1996. Veloce Publishing Plc. ISBN 1 874105 57 X
Triumph Racing Motorcycles in America - Lindsay Brooke. 1996. Motorbooks International. ISBN 0 7603 0174 3

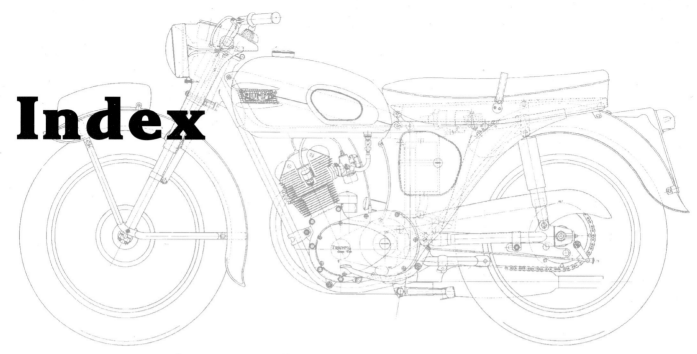

Index

Alternator 11, 13-15, 33, 44, 55, 66-67, 100, 109, 114, 123, 196-97, 199-200, 202-203
Auto-advance unit 14, 55, 66, 100, 201-203

BSA Small Heath factory 18, 20-23, 55, 69-74, 75, 110-11, 113-14, 119-20, 203-204
Bates, USA 30, 99
Battery & battery box 14-16, 50, 55, 62, 74, 77, 100, 108, 114, 199, 201-203
Beasley, Doug 98, 102
Bermuda Cub 58, 89-93, 100, 201
Big end:
    Eccentric crankpin 100, 102
    General 12, 30, 33, 45, 53, 80, 102, 171
    Plain, small 28, 30, 33, 196, 198
    Plain, large 33, 70, 104, 108, 202-203
    Roller, early 13, 33, 85, 196
    Roller, late, caged needles 33-34, 55, 70, 108, 204
Bore & stroke 11, 13, 17, 28, 68, 80, 89-92, 97-98, 100, 102, 104-105, 170
Brakes, hubs & brakeplates 14, 26, 33, 56, 70, 71, 74, 77-78, 83, 96, 104, 106, 113, 176, 197-98, 200, 202
Breathers 13, 15, 30
Build quality 70, 110

CGCIM, France 107-108, 111, 118, 203
Cables 11, 15, 24, 32-33, 55-56, 100, 108, 196, 198, 202
Camshaft 13, 23, 29, 51, 55-56, 61, 64, 68, 79-80, 83, 85, 97-100, 102, 108, 112, 172, 200, 202-204
Capacity changes 80, 82-85, 97-98, 100, 102, 104-105
Carburettor:
    General 13, 23, 45, 50, 56, 67, 77, 80, 98-99, 102, 177
    Amal 32 47, 177, 201-2
    Amal 332 28-29, 46-47, 58, 102, 177, 197
    Amal Concentric 47, 177, 204
    Amal GP or TT 85, 102, 177, 199
    Amal Monobloc 47, 61-62, 64, 105, 108-109, 113, 177, 200-201, 203
    Zenith 46-47, 58, 100, 177, 199-201
Casting numbers 20, 108
Chainguard 43, 48, 73, 106, 113, 198, 201-203
Chains, primary & rear 12-14, 32-33, 44, 48, 67-68, 77, 83, 108, 113, 173, 197-99, 201
Clutch 13, 32, 44-45, 55, 68, 79-80, 99, 173, 196-97, 199-203
Colours 9, 11, 13, 19, 24-25, 28, 56, 62, 70-75, 78, 91-92, 108, 111, 113-15, 118, 178-79, 200-201, 204
Comerfords & Comerfords Cub 104-106
Contact breaker 12, 14, 16, 32, 44-45, 55, 66, 92, 108-109, 197-203
Crankcase 12-13, 19-20, 30, 32, 35, 44-45, 51, 55, 67, 69, 92, 100, 102, 109-110, 196-97, 199, 201-202
Crankcase undershield 47, 60, 64, 108, 202-203
Cylinder barrel 12-13, 28, 30, 44-45, 55, 67, 82, 89, 92, 98, 100, 102, 104, 108, 170, 197, 199-200, 203-204
Cylinder head 11-13, 15, 28, 36, 44-45, 51, 55, 61, 67-68, 85, 97-98, 100, 102, 104, 108, 197-200, 203

Deliveries, delay 17-18, 28-29, 36, 70, 76-77, 119
Domaines, France 111

ERM, Muret, France 109-110
Earls Court Show 9-10, 17, 28, 36, 196
Energy Transfer 17, 33, 55, 61-62, 64, 66-67, 85, 100, 108, 200-203
Engine covers 12-13, 15, 30, 44-45, 51, 55, 62, 67, 100, 196-202
Exhaust system 14, 28, 35, 45-46, 55, 60-62, 64-65, 68, 70-71, 74, 77, 83, 85-86, 100-102, 106, 108, 113-14, 197-204
Exports 10, 17, 22, 32, 47, 56, 64, 67, 71-72, 75, 77, 89, 91-92, 107-11, 113-115, 118, 120, 204

FVRDE, Chertsey 111-13
Fearon, Bob 39, 40-41
Financial year 19, 21, 23
Flywheels and big end assembly 12-13, 28, 33, 51, 80, 89, 92, 98, 100, 104, 196, 198-200, 202-203
Fochj, Italy 36, 38
Footrests 43, 77, 105, 108, 113, 198-99, 201-203
Frames:
    General 14, 19-20, 85, 90, 98, 105, 175
    Plunger type 14-15, 29, 33, 43, 56, 79, 175, 196-97
    Rigid type or conversions 74, 79, 82-83, 86-87, 97, 102
    Swinging fork type 36, 43, 48, 56, 58, 60, 64, 68-71, 73, 80, 97, 100, 102, 105, 108-10, 113, 175, 198, 200-204

France & French machines 10, 21, 23, 46, 56, 64, 72, 100, 107-111, 113, 115, 118, 120, 201-204

Gear indicator 11, 13, 15, 28, 55, 196, 198, 202-203
Gearbox & gearing 11, 13, 15, 26, 29-30, 44-45, 48, 55-57, 60-61, 64-65, 68, 77, 80, 82, 97, 102, 105, 108-110, 173-74, 197-98, 200-204

Handlebars, levers & mirror 34, 74-75, 85, 105, 108, 197, 202-203
Hitchcock, Jock 95-96, 104-105, 196
HT coil 14-15, 66, 100, 108-109, 198

IFVME 113
Ignition timing 16, 32, 55, 66, 98, 100, 109, 125-67, 171, 200
ISDT - International Six Day Trial 43, 58, 95-96, 98, 100, 112-13
Italy & Italian machines 36, 81-82, 85, 98, 102, 120

Japan & Japanese machines 65, 85, 92, 108, 111-21
Johnson Motors, California 29, 62, 64-67, 75-77, 80, 83, 85, 105, 200, 202-203

Keyways 13, 33, 66, 100, 200
Kickstart lever 15, 45, 77, 105-106, 197, 203-204

Lights and lighting 11, 15, 55, 66-67, 72, 74, 106, 108, 113, 196, 200-203
Lubrication system 11-12, 30, 32, 44-45, 48, 51, 53, 55, 68, 70, 77, 80, 83, 104, 108, 113
Lucas 14-15, 33, 40, 55, 67, 74, 109, 196-97, 200-203

Magneto & flywheel magneto 14, 80, 99-101, 199
Main bearings 12-13, 32, 45, 51, 92, 98, 104, 112, 171, 201-202
Marque loyalty 17, 26, 35, 69, 71, 119
Martin, Bill, USA 43, 80, 85-86, 88, 98, 101
Masters, Alec St John 39-41
Meriden factory 10, 14, 17-21, 24, 29-30, 32-33, 36, 39-40, 43-45, 47, 50, 65-71, 74, 77, 82, 91-92, 96, 98, 100-101, 105-106, 108-113, 120, 203-204
Miller, Sammy 66, 68, 105
Minimum amount of metal policy 18, 30, 35-36, 119
MIRA 113
Models derived from the T20 (alphabetical order):
   T20 Bantam 175 74, 120, 204
   T20B Bantam Cub 15, 20-21, 32, 48-49, 56, 69-73, 118, 165, 195, 204
   T20B Pastoral Cub 74
   T20B Super Cub 15, 20-21, 32, 47-49, 56, 69, 71-73, 111, 115, 118, 120, 167, 195, 204
   T20B Tarbuk 72-74, 204
   T20C Competition Cub 43, 58-61, 64, 66, 80, 100, 118, 143, 198-200
   T20CA 56, 58, 60-61, 66, 78, 145, 198-200
   T20CB 58, 61, 200
   T20J Junior Cub 47-48, 56-58, 78, 133, 135, 137, 139, 199-201
   T20MWD French Army 22, 56, 64, 72, 107-11, 120, 163, 201-203
   T20P Police 45, 48, 114-15, 118, 203-204
   T20S Sports 22-23, 44, 47, 56, 60-62, 64, 66, 77, 81-82, 102, 111-13, 120, 147, 200-201
   T20SA Sports, America 62
   T20SC Sports, Competition 56, 64-67, 71, 78, 120, 157, 202-204
   T20SH Sports, Home 23-24, 48, 62, 64, 66, 71-73, 155, 194-95, 201-204
   T20SL Sports, Lights 61-63, 66, 151, 194, 201-202
   T20SM & T20M Mountain Cub 21-24, 32, 56, 64-66, 72, 75, 78, 104-106, 108, 111, 115, 118, 161, 202-204
   T20SR Sports, Road 23, 56, 62, 64-67, 78, 157, 202-204
   T20SS Street Scrambler 23, 56, 63-66, 78, 92, 153, 202-203
   T20T Trials 22, 63-64, 66, 108, 110, 112-13, 149, 200-203
   T20W Woods 22-23, 63, 66, 147, 201
   T20WD War Dept 22, 64, 66, 108, 110-13, 201-203
   TR20 Trials 23, 56, 63-66, 100, 104-105, 108, 159, 202-203
   TS20 Scrambler 56, 63-64, 66, 159, 202-204
Mudguards 11, 28, 36, 70, 74, 77, 105, 196, 198-99, 201-203
Mute 45, 108, 115, 200

Nacelle 11, 15, 28, 32, 50, 55, 60, 75, 197-99, 202

Oil filter 44-45, 80, 102, 199-201
Oil pump, pressure, relief, leaks 12-13, 29-30, 34, 45, 48, 53, 55, 70, 80, 102, 108-10, 113, 119-20, 172-73, 196-97, 201-202, 204
Oil seals:
   Clutch 13, 30, 109-10, 197, 199
   Engine sprocket 13, 196
   Front forks 44, 198-99
   High gear 13, 30, 44, 196, 199, 204
   Side points 55
Oil tank 12-13, 30, 50, 60, 67, 70-71, 74, 100, 102, 104-106, 108, 196-99, 202, 204
Other countries:
   Africa 28, 58, 111, 113-14, 118, 120-21, 181-93, 204
   Americas (except USA) 28, 58, 110, 113, 120-21, 181-93, 204
   Australasia 21, 28, 36, 74-78, 120-21, 181-93, 196
   Europe 9, 21, 28, 36, 78, 102, 107-11, 113, 120-21, 181-93, 196, 200-202
   Middle & Far East 28, 36, 72, 113, 118, 120-21, 181-93, 203-204

Performance & parts 9-10, 14-15, 18, 26, 28, 34, 36, 39-40, 45, 47, 56, 58, 61-62, 67-71, 78-79, 82-83, 85, 91, 95, 97-100, 102, 104, 113, 119, 194-95
Petrol tank:
   Badges 11, 48-49, 69-72, 74, 107, 198-99, 201, 204
   Internal baffles 48, 200
   Types of tank, etc. 33, 36, 48, 50, 58, 64, 70-71, 73-74, 100, 104-105, 113-14, 175, 197-99, 201-204
Peugeot, France 107-108, 111
Peyron, Guy, France 107-108, 110
Pistons and piston rings 13, 28-29, 45, 56, 60-61, 63-64, 79-80, 83, 85, 97-98, 100, 102, 104, 108, 110, 170, 197-98, 200, 203-204
Police 19, 45, 48, 92, 113-18, 202-203
Purchase tax 17, 20-21, 77

Racing:
   Drag-racing 81, 85, 94, 101
   Enduro 81, 86, 94-95
   Grass track 94, 100, 102
   Ice racing 94-95
   Road racing 56, 81-82, 85-86, 94, 96-99, 102, 104
   Scrambles 81, 82, 94, 100
   Short track 56, 81-84, 94, 102, 104
   Speedway 56-57, 81, 102
Radio equipment 48, 114-15, 202-203
Records (factory) 19-23, 36, 40, 58, 60-61, 71-72, 74-75, 77, 105, 108-109, 111, 114-15, 118, 120, 196, 200-203

Rectifier 14, 16, 33, 108, 197, 200, 202
Register, Tiger Cub & Terrier 22-23, 58, 72, 106, 109, 111, 122
Road tests 20, 28-29, 33, 70-71, 77, 194-95
Rocker box and covers 13, 15, 30, 55, 80, 83, 104, 196, 199, 202

Sangster, Jack 18
SSDT - Scottish Six Day Trial 43, 96-97, 100-101, 105, 201
Seat or saddle 11, 28, 50, 74-75, 77, 91, 105, 108, 113-15, 196-98, 200-203
Serial number 19-23, 70, 92, 100, 108-109, 111, 115, 168-69, 203-204
Side panels 48-50, 68, 199, 203
Skew gear 12, 16, 32, 34, 55, 66, 100, 200, 202-203
Sludge trap 12, 51, 196, 203
Sno-Go 104-105
Speed records 80, 85, 86, 101
Speedometer and drive 11, 28, 55-56, 62, 106, 113, 198, 200-202
Sprockets:
    Engine and clutch 32, 174, 197, 199
    Gearbox 15, 26, 28, 44-45, 60, 80, 99, 105-106, 108, 109-10, 113, 174, 198-99, 201
    Rear wheel 80, 97, 99, 102, 105, 108, 110, 113, 174, 198, 201
Stands, centre and side 29, 33, 100, 110, 113-14, 175, 196-98, 200-201, 203
Steering lock 43-44, 199
Sturgeon, Harry 69, 110
Summers, Harry 10-11
Suspension:
    front 14, 36, 43-44, 48, 60-61, 64, 73-74, 77, 79, 85, 97-98, 100, 104, 106, 113, 174, 198-200, 203
    rear 14-15, 26, 33, 36, 43-44, 48, 60, 73, 79-80, 82-83, 85, 97-98, 100, 102, 175, 197-98, 200, 202-203
Switches 11, 16, 32, 55, 67, 74, 196, 202-203

Tachometer 55, 60-62, 96, 102, 199-202, 204
Tait, Percy 102, 113
Tensioner, primary chain 12
Terrot, France 107-108, 201-202
TIB4 102
Timing pinion 12, 16, 34, 55
Trials and trials machines 43, 47-48, 81, 94-106, 108, 111, 122, 201-202

Triumph Corporation, Baltimore 30, 40, 62, 64-65, 67, 75-77, 83-84, 102, 104, 200-204
Truslove, Stan 109
Tyres 14, 28, 48, 55-56, 58, 60-61, 65, 74, 77, 85, 96, 108-109, 197, 199-200, 203-204
Turner, Edward 9-12, 17-18, 24, 29, 35-36, 39-41, 65, 67-69, 89-90, 98, 119-20

Unit construction 10-11, 13, 69
United Kingdom 10, 24, 28, 36, 58, 65, 77-78, 94-95, 102, 113-14, 120, 180-81, 203-204
United States of America 10, 17, 21, 24, 28, 30, 32-33, 40, 43-44, 47, 50, 56-58, 60-62, 64-67, 69, 75-88, 92, 94, 97-102, 104-105, 120, 191-92, 196, 198-203

Vale, Henry 43, 113
Valves and valve gear 11-13, 20, 35, 45, 51, 61, 68, 77, 79, 80, 83, 85, 92, 97, 99, 100, 102, 104, 108, 112, 172, 197-201, 203

Wall of Death 94
War Office 64, 111-13
Warranty 70, 110
Webco, USA 30, 80, 85
Wet-sumping 30, 201
Wheels & wheel bearings 14, 24, 36, 48, 55-56, 58, 70-71, 74, 77, 83, 85-86, 91-92, 105, 108, 113, 176, 197-200
Wickes, Jack 10-11, 14, 80
Wipac 14-17, 74, 196, 204
Wiring 15-16, 32, 55, 67, 109, 113-14, 122-23, 196-97, 202
Woods kit 64, 102
Works trials riders:
    Alves, PH (Jimmy) 96-97
    England, Paul 102, 105
    Farley, Gordon 102, 105-106
    Fisher, George 43, 100, 102
    Heanes, Ken 43, 58, 100, 102
    Peplow, Roy 100-102, 105
    Ratcliffe, Artie 100-101, 102
    Rathmell, Malcolm 102, 105
    Sayer, Ray 101-103
    Others (Gordon Blakeway, Scott Ellis, John Giles, Dave Thorpe) 102

Visit Veloce on the Web - www.veloce.co.uk